SATELLITE BROADCASTING SYSTEMS
Planning and Design

ELLIS HORWOOD SERIES IN ELECTRICAL AND ELECTRONIC ENGINEERING

Series Editors: PETER BRANDON, Professor of Electrical and Electronic Engineering, University of Cambridge
P. R. ADBY, Department of Electronic and Electrical Engineering, University of London King's College

APPLIED CIRCUIT THEORY: Matrix and Computer Methods
P. R. ADBY, University of London Kings's College

BOND GRAPHS FOR MODELLING ENGINEERING SYSTEMS
ALAN BLUNDELL, Coventry (Lanchester) Polytechnic

NOISE IN ELECTRONIC DEVICES AND SYSTEMS
M. J. BUCKINGHAM, Royal Aircraft Establishment, Farnborough, Hampshire

DIGITAL AND MICROPROCESSOR ENGINEER
S. J. CAHILL, Ulster Polytechnic

DIFFRACTION THEORY AND ANTENNAS
R. H. CLARKE and JOHN BROWN, Imperial College of Science and Technology, University of London

INTRODUCTION TO MICROCOMPUTER ENGINEERING
D. A. FRASER, *et al.,* Chelsea College, University of London

COMPUTER-AIDED DESIGN IN ELECTRO-MAGNETICS
E. M. FREEMAN and D. A. LOWTHER, University of London, and P. P. SYLVESTER, McGill University, Montreal

NON-LINEAR AND PARAMETRIC CIRCUITS: Principles, Theory and Applications
F. KOURIL and K. VRBA, University of Brno, Czechoslovakia

ELEMENTARY ELECTRIC POWER AND MACHINES
P. G. McLAREN, University Engineering Department, Cambridge

MICROWAVE PHASE MODULATIONS
T. MORAWSKI and J. MODELSKI, Warsaw Technical University

CONTROL SYSTEMS WITH TIME DELAYS
A. W. OLBROT, University of Warsaw

POST-DIGITAL ELECTRONICS
F. R. PETTIT, Computing Teaching Centre, Oxford

HANDBOOK OF ELECTRONIC CIRCUITS
G. J. SCOLES, formerly English Electric Valve Company, Chelmsford, Essex

HANDBOOK OF RECTIFIER CIRCUITS
G. J. SCOLES, formerly English Electric Valve Company, Chelmsford, Essex

PRINCIPLES OF COMPUTER COMMUNICATION NETWORK DESIGN
J. SEIDLER, Technical University of Gdansk, Poland

SATELLITE BROADCASTING SYSTEMS

Planning and Design

J. N. SLATER
Senior Engineering Information Officer
Independent Broadcasting Authority, Winchester

and

L. A. TRINOGGA, M.Sc., Ph.D. (CNAA)
Reader in Electrical Engineering, Leeds Polytechnic

ELLIS HORWOOD LIMITED
Publishers · Chichester

Halsted Press: a division of
JOHN WILEY & SONS
New York · Chichester · Brisbane · Toronto

First published in 1985 by

ELLIS HORWOOD LIMITED
Market Cross House, Cooper Street, Chichester, West Sussex, PO19 1EB, England

The publisher's colophon is reproduced from James Gillison's drawing of the ancient Market Cross, Chichester.

Distributors:

Australia, New Zealand, South-east Asia:
Jacaranda-Wiley Ltd., Jacaranda Press,
JOHN WILEY & SONS INC.,
G.P.O. Box 859, Brisbane, Queensland 4001, Australia

Canada:
JOHN WILEY & SONS CANADA LIMITED
22 Worcester Road, Rexdale, Ontario, Canada.

Europe, Africa:
JOHN WILEY & SONS LIMITED
Baffins Lane, Chichester, West Sussex, England.

North and South America and the rest of the world:
Halsted Press: a division of
JOHN WILEY & SONS
605 Third Avenue, New York, N.Y. 10158 U.S.A.

British Library Cataloguing in Publication Data
Slater, J.N.
Satellite broadcasting systems: planning and design. –
(Ellis Horwood series in electronic and communication engineering)
1. Artificial satellites in telecommunication
2. Broadcasting
I. Title II. Trinogga, L.A.
621.38'0422 TK5104

Library of Congress Card No. 85–14031

ISBN 0–85312–864–2 (Ellis Horwood Limited)
ISBN 0–470–20217–3 (Halsted Press)

Printed in Great Britain by R.J. Acfords, Chichester

Table of Contents

Authors' Preface

'A book for all reasons' might well have been the alternative title for this collection of information about all aspects of satellite broadcasting, for its authors wanted, right from the start of their project, to produce a book that would contain something for everybody.

The engineering student wanting to go into the detail of any facet of the subject should be able to find enough information to meet his needs, and the book may well become a standard textbook for undergraduate engineering courses.

At the same time the book is laid out in such a way as to allow the rather less dedicated reader, who needs only a general outline of each aspect of satellite broadcasting, to find all that he wants without difficulty. Each chapter starts with a very readable and easily assimilated section giving the basic information about the particular subject under consideration, and this is then followed, when required, by a mathematical treatment of the subject.

Reflecting the up-to-date nature of its subject, the book contains some specially-developed computer programs which should prove interesting and helpful to the evergrowing number of people who have access to a microcomputer.

From the basic physics of satellite orbits to more detailed calculations of amplifier parameters, this book contains a lot of theoretical instruction, but this is balanced throughout by sound practical information about everything from the choice of rocket launchers to the design of satellite receiving equipment.

The authors hope that whether the reader merely dips into various parts of the book or whether he uses it for detailed study, he will find what he is looking for, and that the book will form a valuable resource and a useful compendium of knowledge about this fascinating subject.

No book as wide-ranging as this one can honestly claim to have started 'from scratch' on all topics, and the authors gratefully acknowledge their indebtedness to those who have gone before. In particular, *Technical Review* 11, published by the Independent Broadcasting Authority and now out of print, provided much valuable source material for the sections on fundamentals, and thanks are due to Alfred Witham, Ken Hunt and Fred Wise, who were responsible for much of the original material. The authors are grateful to the Director of Engineering of the IBA for his permission to quote from this document.

The views expressed in this book are those of the authors, and do not necessarily represent the views of the Independent Broadcasting Authority.

1

Satellites – a conjunction of scientific disciplines

Unlike many scientific subjects, which can only be of interest to a few privileged cognoscenti with a particular love of some obscure branch of knowledge, satellite broadcasting holds a peculiar fascination for a wide range of people. These range from the viewer 'in the street' who has been led to expect satellites to bring him an infinite choice of television programmes on demand, through the physicists and mathematicians who take delight in calculating the precise orbits of the satellites, to the engineers whose practical and constructional skills have made the satellites feasible.

All branches of engineering are vital to the successful operation of satellite broadcasting, and the quest for better ways of making use of satellites has led to frontiers being moved forward in many fields of engineering and materials science. It will readily be appreciated that electronics engineers and rocketry and ballistics experts make a significant contribution to satellites, but it is not always so well recognised that considerable civil engineering skills are also needed – how else could we ensure the precision performance of huge receiving dishes such as those used by British Telecom at Goonhilly in Cornwall (Fig. 1).

Similarly, the contributions of mechanical engineers are vital to the proper functioning of spin-stabilised satellites and to the correct deployment of the beautifully engineered solar panels which smoothly emerge from the confined space of the launch vehicle to expand into space like the wings of a butterfly coming out of its chrysalis (Fig. 2).

A full understanding of satellite broadcasting will, then, involve a study of many different subjects, and we shall be looking in detail at these later on. Let us start, though, at the beginning, with the first published ideas on satellite broadcasting, which came from the well-known American writer of science fact and fiction, Arthur C Clarke (Fig. 3).

Fig. 1 – Photograph of Goonhilly dish (photo – British Telecom).

1.1 HISTORICAL

In an article published in *Wireless World* in 1945 [7] Clarke foresaw that it would be possible to provide complete radio coverage of the world from just three satellites, provided that they could be precisely placed in a particular path known as a geostationary orbit (Fig. 4).

The important feature about the geostationary orbit is that the rotational speed of a satellite in this orbit is equal to that of the earth, so that although the satellite will be rushing along at around 11 200 km per hour, it will appear to any observer on earth as though it is hanging motionless in the sky. Although Clarke envisaged large manned space-stations in this geostationary orbit, so far this has not happened, and more importance is now given to the fact that if

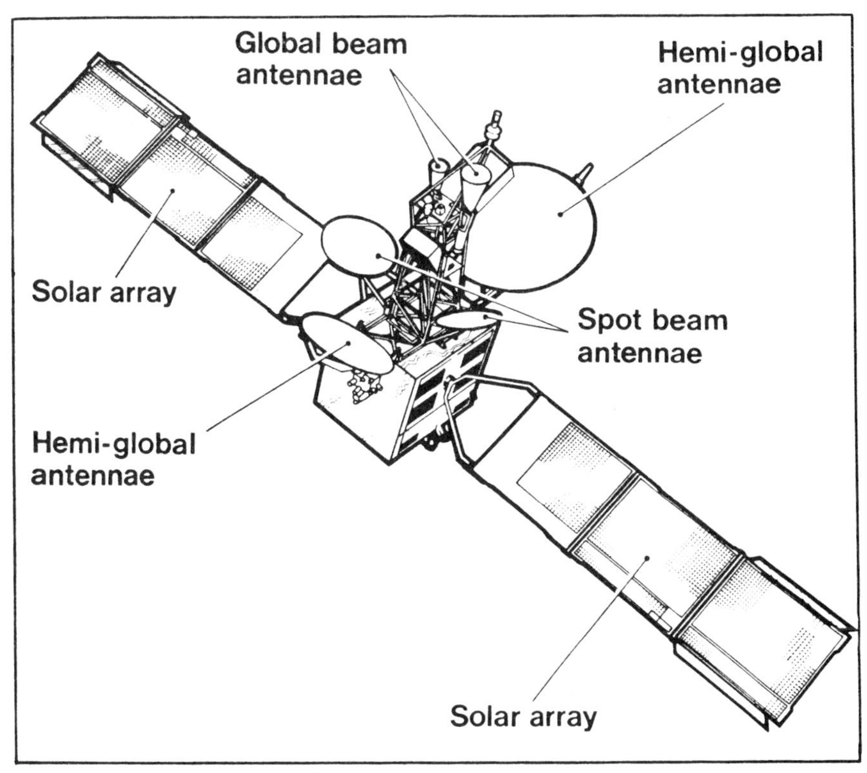

Intelsat V: explanatory diagram showing principal external components.

Fig. 2 – Intelsat V: explanatory diagram showing principal external components.

this orbit is used for broadcasting satellites, viewers on earth will be able to use fixed receiving aerials, since the satellites appear to be fixed in position, rather than the immensely complex tracking aerials that would be needed to continuously follow broadcasts from satellites in other orbits around the earth. The geostationary orbit is, of course, merely a special case of the more general elliptical orbit. The mathematics of the ideal geostationary orbit are straightforward enough, bearing in mind that a satellite in this orbit will, by definition, need to have the same period of rotation as that of the earth, and the orbit will need to be circular and in the plane of the equator.

1.2 THE EARTH'S GEOMETRY

The earth is approximately spherical, with a radius of 6376.77 km, and it rotates upon its axis approximately once every 24 hours. Because the earth not only rotates on its own axis but also rotates around the sun once per year, the

Fig. 3 – Arthur C. Clarke (photo – *Wireless World*).

accurate measurement of the length of the day is fairly complicated. The siderial day, that is a day measured by reference to a distant star, differs in length from a solar day, that is a day measured by reference to the sun. The duration of a year is 365.25 days, but during that time the earth actually rotates 366.25 times around its axis. From this we can see that the siderial

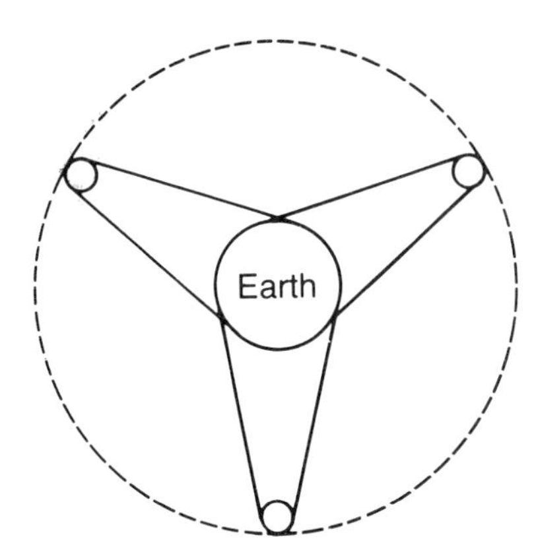

Fig. 4 – Complete coverage of earth's surface from three satellites.

period of rotation is 24 hours × 365.25 divided by 366.25, and this works out to be 23 hours 56 minutes and 4.1 seconds.

The plane which contains the earth and the sun is known as the ecliptic, and the axis of the earth is not in fact perpendicular to the plane of the ecliptic but is tilted at an angle of 23° 27′, an angle known as the obliquity of the ecliptic (Fig. 5).

Gravity

When considering the motion of a satellite around the earth we need to consider in some detail the gravitational effects of the earth. If the earth were truly spherical and assumed to be of uniform density, then its gravitational attraction for a satellite of mass m would be given by the equation mgR_e^2/r^2 where g is the acceleration due to gravity at the earth's surface, R_e is the radius of the earth, and r is the distance of the satellite from the earth's centre. The earth is not in fact a perfect sphere, its radius at the equator being 6376.77 km whilst its radius at the poles is about 6355 km, as shown in Fig. 6.

Because our geostationary satellites will orbit in the plane of the equator we will use the equatorial radius for our calculations, although it is important to note that the non-spherical shape of the earth will introduce the need for a correction factor. The gravitational attraction is in fact modified by a factor $1 + 1.64 \times 10^{-3} (R_e/r)^2$, which makes a difference of around 0.0015% in the radius of the geostationary orbit. Although this correction would need to be taken account of in real life satellite calculations, it will be ignored for our purposes.

The fact that the earth is rotating, naturally means that its surface will have a significant velocity component. Figure 6 shows how we may calculate this. In Fig. 6, s is the radius from the axis of rotation to a point on latitude 0°

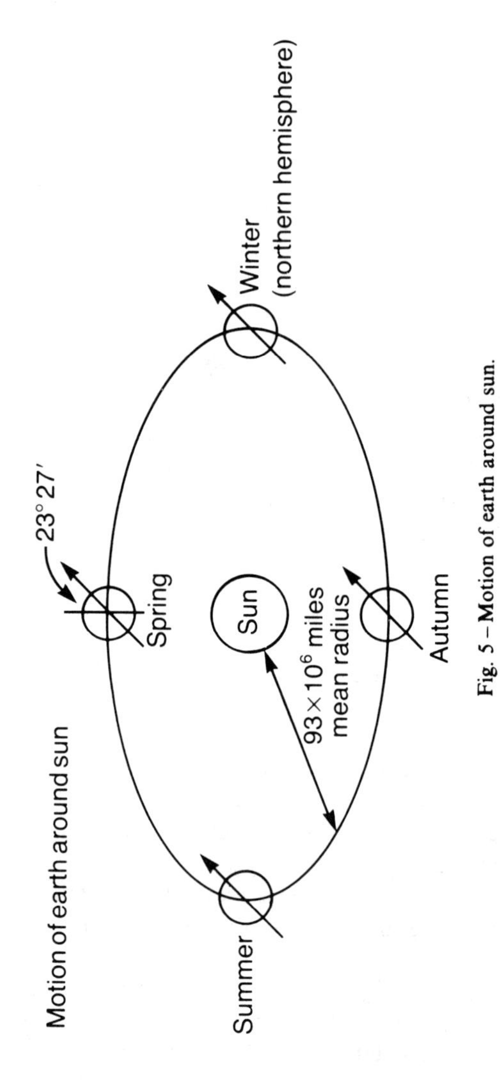

Fig. 5 – Motion of earth around sun.

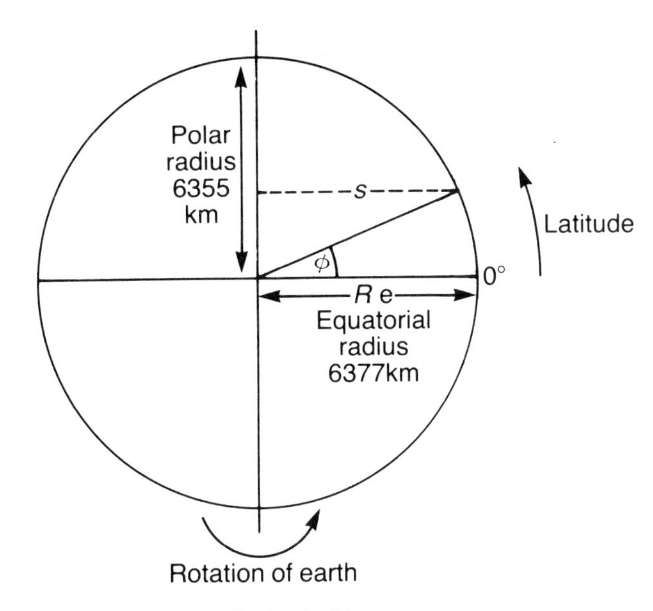

Fig. 6 – Earth's geometry.

R_e is the radius of the earth, assumed spherical. Hence $s = R_e \cos \phi$.

We stated earlier that the siderial period of rotation of the earth is 23 hours 56 minutes 4.1 seconds, and from this and the knowledge that R_e equals 6376.77 km we work out that the surface velocity (m/s) at any particular latitude ϕ is $465 \cos \phi$. We shall see later that this surface velocity is put to good use when satellites are launched.

1.3 ORBITAL CALCULATIONS

Any satellite travelling in a fixed circular orbit is effectively in a condition of equilibrium, and we can therefore consider the situation to be shown in Fig. 7, where the centrifugal force F_c owing to the satellite's motion around the earth, which will tend to push the satellite away from the earth, is counterbalanced by the gravitational force of the earth F_g, which is trying to pull the satellite back towards the earth.

$$F_c = F_g \qquad (1.1)$$

or

$$\frac{m_s v^2}{r+h} = \frac{m_s G m_e}{(r+h)^2} \qquad (1.2)$$

where m_s = mass of satellite (kg)
m_e = mass of earth, 5.977×10^{24} kg
v = velocity of satellite (m/s)

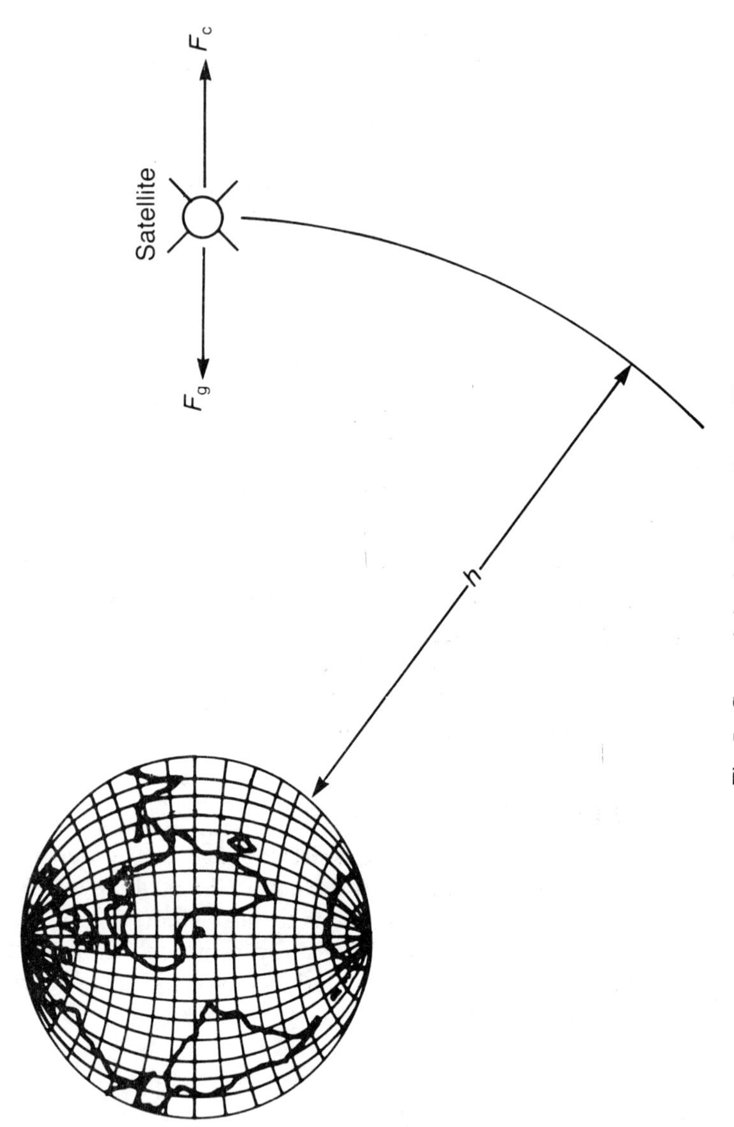

Fig. 7 – Counterbalancing forces of a satellite.

h = height of satellite above earth (m)
r = radius of earth, approximately 6377 km
G = constant of gravitation, 6.67×10^{-11} m^3 kg^{-1} s^{-2}
F_g = gravitational force (N)
F_c = centrifugal force
[kg m/s^2 = 9.81 N]

From equation (1.2) the satellite velocity is thus given by

$$v = \sqrt{\frac{G m_e}{r+h}} \tag{1.3}$$

The time taken for the satellite to make one complete rotation around the earth, its orbital period, is therefore

$$T = \frac{2\pi(r+h)}{v} = 2\pi\sqrt{\frac{(r+h)^3}{Gm_e}}. \tag{1.4}$$

We have defined a geostationary orbit as one whose period is the same as that of the earth's rotation around its axis, i.e. 23 hours 56 minutes 4.1 seconds, so if we substitute this value for T in the equations above we can calculate that the radius of the geostationary orbit $(r+h)$ is 42 157 km, so that the geostationary satellite will be at a height of 35 779 km above the equator. Similar calculations can be performed to obtain each of the parameters of an ideal geostationary orbit, and these are listed in Table 1.1.

Table 1.1 Geostationary orbit parameters

satellite period T	23 h 56 mins 4 s = 86 164 s
radius of circular earth r	6 377 km
radius of orbit $d = r + h$	42 157 km
equator inclination	0°
velocity of satellite v	3.074 km/s
eccentricity, circular orbit	0
altitude of satellite h	35 779 km

REFERENCES AND BIBLIOGRAPHY

[1] W. Stoesser, Direct broadcast satellite systems. *Microwave Journal*, **25,** No. 8, Aug. 1982, p. 99.
[2] Issue devoted to satellite communications *Microwave Journal*, **25,** No. 1, Jan. 1982.
[3] Special issue on satellite communication networks *Proc. IEE*, Nov. 1984.
[4] *IBA Technical Review* No. 11.
[5] *IBA Technical Review* No. 18.
[6] *IBA Broadcast Engineering Notes* No. 2 (out of print).
[7] Arthur C. Clarke, "Extra-terrestrial relays"—*Wireless World* Oct. 1945.

2

Satellite launching

2.1 BASIC TECHNIQUES

Putting satellites into geostationary orbits is far from easy, and it was some seven years after the launch of the first earth-orbiting satellites that engineers from the Hughes Aircraft Corporation finally managed to come up with a usable method of manoeuvring a satellite into geostationary orbit. The method which has now been refined and is in regular use, is shown in Fig. 8.

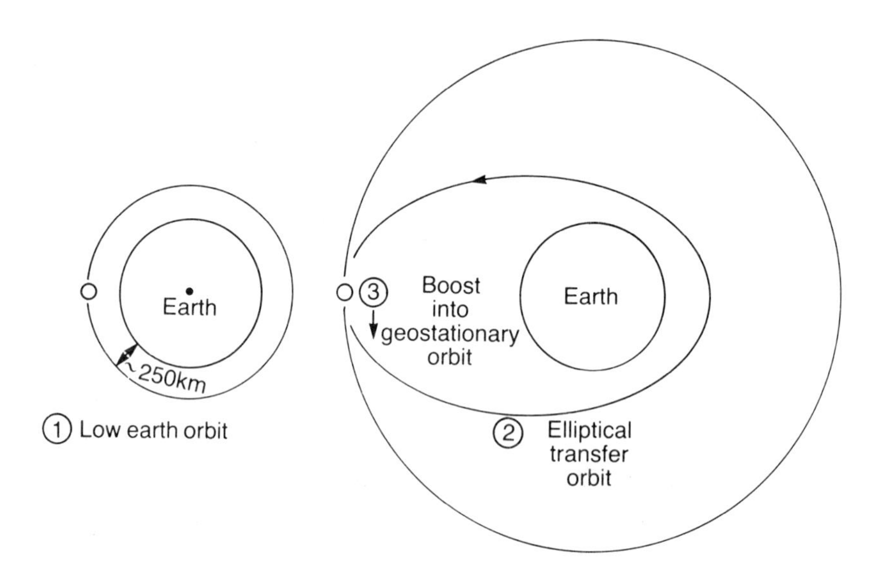

Fig. 8 – How a geostationary orbit is achieved.

The satellite is launched, preferably from an equatorial site so as to make the most of the helping hand given by the surface velocity of the earth, and is put into a low earth orbit, usually around 250–300 km above the earth's surface. Once it is established satisfactorily in this low orbit its velocity is increased to take it into an extremely ellipitical orbit, whose perigee (nearest distance from the earth) is around 270 km, but whose apogee (furthest distance from the earth) is around 36 000 km, the eventual height of the geostationary orbit. When the satellite has completed a couple of these so-called transfer orbits and the scientists are happy that all is functioning properly, a rocket motor is fired when the satellite reaches the apogee position, and this sharply increases the velocity of the satellite to 3.074 km/s which is the velocity needed to take the satellite into circular geostationary orbit and keep it there. The so-called 'apogee-boost' motor that finally puts the satellite into its correct orbit has frequently been of solid-fuel design, since it is only designed to burn once in the whole lifetime of the satellite; but more recent designs have found it more economical to arrange to use the same hydrazine gas fuel that is normally used to power the satellite's small positioning rockets which are used to make final adjustments to bring the satellite precisely into its orbital position.

It is very important that a geostationary satellite designed to broadcast to viewers with fixed aerials on earth is kept accurately in its orbital position, since if it is not the satellite will drift with respect to the location of an observer on earth. The drift will be westwards (i.e. against the rotation of the earth) if its orbital radius is too large, and vice versa. As we shall show later (problem 1, page 70) a drift of one degree per day corresponds to a change in orbital distance of 80 km. This sort of error is easily coped with by tracking dishes on earth, but could significantly affect the quality of television pictures received by viewers with fixed dishes. Another type of problem arises if the satellite is slightly inclined to its proper position, or in other words, if its orbit does not lie completely in the equatorial plane. Since the satellite's orbital plane will be constant, but the earth will be rotating, any observer on earth will see that the satellite appears to oscillate, carrying out a figure-of-eight movement once a day. It is normally possible to position a satellite within about 30 metres of its ideal spot, and this minimises the corrective positioning that will need to be carried out during the satellite's operational life.

One of the major factors affecting the longevity of an operational satellite is the amount of fuel that can be carried to feed the positioning rockets, so it is obvious that the more accurate the initial launch positioning is, the more fuel will be left to cover the many minor adjustments that will be needed during the lifetime of the satellite. A recent example of how this effect can be used 'in reverse', so to speak, was when the American Tracking and Data Relay satellite (TDRSS-A) launched from the Challenger Space Shuttle failed to achieve its proper geostationary orbit on the initial launch. Judicious use of the satellite's positioning rockets, which were fired in carefully controlled short bursts over a period of several weeks, eventually brought the satellite into its geostationary orbit, so saving many tens of thousands of dollars which would have been wasted if the satellite had not been able to obtain its proper orbit. Although much work was done to calculate the best way of making the most effective use of each hydrazine 'burn', the real cost of the exercise will be that the amount of fuel left for satellite positioning will be much reduced, which could have a deleterious effect on its operational lifetime.

The proposed UK satellite UNISAT, which has a design life of 7 years, has deliberately been given fuel tanks that should provide enough fuel for 10 years of life, just to ensure that fuel shortage will not lead to unexpected failure. Up to the present time most satellite designs have erred on the side of over-engineering so as to ensure that the generally stringent contract specifications are met or exceeded, and it is not uncommon for such specifications to demand that a satellite shall be capable of providing a certain transmitter output power at the end of its design life of 5 or even 7 years. The Orbital Test Satellite, OTS–2, is a good example of a satellite which has provided far more than its designers originally envisaged. Launched in May 1978 with a designed life of 4 years, this satellite was still giving excellent service when it was eventually switched off during 1984 to allow its more up-to-date replacement, ECS, to take over. In contrast, though, its predecessor OTS–1 blew up on the launch pad, and several transponders on the Japanese medium-power satellite BSE–1 (Broadcast Satellite Experiment) and on its successor 'Yuri', failed prematurely, so there can be no guarantees of satellite lifetimes, and satellite insurance premiums are likely to remain high for the foreseable future in spite of all that has been learned from the 'zero-faults' technology that the Americans pioneered for the moon-shots in the sixties.

Fig. 9 – Orbital test satellite (OTS).

Atmospheric drag

The effects of the atmosphere upon a satellite vary considerably with the height of the satellite orbit above the earth. Table 2.1 shows pressures and temperatures to be expected at various altitudes.

Table 2.1 Temperature and pressure

Altitude (km)	Pressure (mm of mercury)	Temperature (K)
200	10^{-6}	700
400	10^{-8}	1500
600	10^{-9}	2200
800	2×10^{-10}	3000
1000	10^{-10}	4000

At low altitudes atmospheric drag is a significant factor on the orbit of any satellite travelling through it. Table 2.2 shows some calculations which have been made to express the effects of this drag in terms of how much energy is lost by the satellites during each complete orbit of the earth.

The table shows that satellites orbiting at only a few hundred kilometres above the earth must have a fairly short life, but the effects of the atmosphere on geostationary satellites which, as we have seen, are at a much greater distance from the earth, can be regarded as negligible.

Table 2.2 Typical variation of energy loss with height of orbit

Orbit height (km)	Energy loss (%)
120	1.4
160	0.2
200	0.04
240	0.01

The effects of radiation on a satellite in geostationary orbit

The earth is constantly subjected to radiation made up of high energy electrons and protons which are trapped by the earth's magnetic field. Over the years it has been shown that an increase in the intensity of this radiation takes place after the occurrence of nuclear storms seen on the surface of the sun, and so it seems reasonable to assume that some of these particles emanate from the sun. The existence of belts of radiation around the earth, known as Van Allen belts, was confirmed by a satellite called Explorer 1 in 1958. It was found that the radiation tends to be concentrated over the equatorial region of the earth, and that there is a very high density of very energetic particles at around 16 000 kilometres above the earth's surface. The geostationary orbit is far enough away from the earth's surface for the radiation not to pose any insuperable engineering problems, but any satellite being launched into this orbit must pass through the Van Allen belt, and this factor must be taken into account in the satellite's design.

Temperature effects

The temperature of a satellite in geostationary orbit will be dependent only upon the amount of radiation which it receives from the sun at any particular time, and it will therefore be warm when exposed to the sun, but will become extremely cold during those periods when it is eclipsed by the earth, a topic which we shall consider later. To get some idea of the temperature likely to occur on our geostationary satellite it is necessary to determine the energy density radiated by the sun in the vicinity of the earth, and this can be shown to be about 1.39 kW/m^2.

Once the satellite has reached thermal equilibrium, the energy radiated by the satellite is equal to that being absorbed by the satellite from the sun's radiation falling upon it. It is generally true that the energy radiated from a body is proportional to its total surface area, whereas the energy intercepted from the sun's radiation is proportional to its cross-sectional area, i.e. the effective area presented to the sun. It is obvious, therefore, that the equilibrium temperature will depend to a great extent upon the shape of the satellite and on its orientation towards the sun. The temperature will be modified by the surface finish of the satellite and also by the heat that would need to be dissipated from its electrical equipment. At the time of writing, much research is being carried out to try to discover the reasons for the failure in 1984 of the high-power travelling wave tubes used by the Japanese 'Yuri' satellite. The problems of the effective dissipation of the large amounts of heat produced by such high-power devices are yet to be solved.

2.2 SATELLITE LAUNCH VEHICLES

If we are to achieve geostationary orbits for our satellites, then we must have some means of accelerating them to the speeds necessary for those orbits to be maintained, and the only known practical way of doing this is to use some form of jet motor or rocket. Because of the absence of air in the upper atmosphere the only motor which can be considered is that type in which all the fuel, oxidant as well as the combustible (usually hydrogen), is carried, and this implies the use of a rocket.

Although we tend to think of this application of rocketry as a fairly recent idea, the Russians put forward the idea of a liquid oxygen/hydrocarbon fuelled rocket motor in a paper 'Exploration of universal space by jet propulsion devices' in 1903. Goddard, in the USA, developed rocket technology around 1914, and by the 1930s more or less satisfactory rocket engines using liquid fuel were available in the USA, Russia, and, of course, in

Germany. It was in the latter country that work under such people as Oberth and von Braun brought the technology to the point where production of reliable motors was quite practicable, with the V-2 rocket being perhaps the most famous.

During the mid-1950s the Americans made much publicity out of the launching of their Inter Continental Ballistic Missiles, but it was the Russians who used these techniques to put the first artificial satellite into orbit on 4 October 1957, and from that time onwards techniques of putting machines and men into space have rapidly advanced.

A high point in the technology was reached in July 1969 when an American Saturn V rocket, burning fuel at the rate of about 12 tonnes/s, lifted its 3000 tonnes off the earth, and for the first time sent men to the moon.

At the present time the western world has the choice of two principal means of launching geostationary satellites, the European 'Ariane' or the American space shuttle. It is also possible to make use of various American expendable launch vehicles such as the McDonnell Douglas Delta, the Titan 34D, and the General Dynamics convair Atlas/Centaur rockets, which have a history of successful launch flights. Until recently it was felt that the re-usable shuttle would cause these launch vehicles to be phased out, but the current boom in demand for satellite launchings, particularly for telecommunication purposes, and a re-examination of the true costs of shuttle launches, has led to a rethink. Several expendable launch vehicles are shown in Fig. 10.

The American government is now actively encouraging the use of expendable launchers like those mentioned above, and there may even be some element of financial subsidy, since the government is expected to absorb the research and development costs of these vehicles, which were initially developed with military purposes in mind.

The first commercial European launch vehicle, Ariane, had an unfortunate early history, with a high percentage of launch failures due to a number of different faults in the equipment. Ariane had been designed over a period of about six years before its initial launch at the end of 1979. Right from the start the European Space Agency made no secret of the fact that they intended Ariane to be a strong competitor for the United States shuttle, which was due to become operational in the early 1980s. Ariane is a three-stage launcher, built on the classical principles of rocketry with the third stage being driven by a highly advanced cryogenic motor, the HM7, which develops sufficient thrust to put two satellites directly into the highly elliptical transfer orbit, as was demonstrated in 1983 when the sixth Ariane launch sent the European Communications satellite ECS and Oscar 10, an amateur radio satellite into geostationary orbit.

Plans are well advanced for future developments of Ariane which will allow much greater payloads to be carried into orbit. By increasing the thrust of the Viking engines used in the first and second stage of the rockets, and by

Fig. 10 – Expendable launch vehicles.

increasing the amount of cryogenic propellant in the third stage by about 25 %, Ariane–2 should be capable of carrying well over 2000 kg of payload, compared with Ariane–1's current capability of around 1200 kg. Ariane–3 is similar, but has two solid-fuel booster rockets strapped on to the first stage which increase the maximum payload to around 2600 kg. Not content with this, Ariane designers have firm plans for a 'stretched' version, Ariane–4,

Fig. 11 – Atlas rocket.

which, with the aid of four strap-on liquid boosters will permit loads of up to 4300 kg to be placed into the transfer orbit. The Ariane–4 series of launchers should be ready by 1986 to meet the demand for the launch of several heavy direct broadcast satellites (DBS) which are planned to go into operation in that year. The whole concept of Ariane–4 is to provide a flexible package with variations in payload capacity being met by the addition of solid or liquid booster rockets as required by a particular customer. Arianspace is currently quoting a price of around $30 million per satellite, for a dual satellite launch, but one of the major objectives of future Ariane designs will be to reduce the cost per kilogram of payload considerably.

Already the design of an Ariane–5 is being discussed, and the engineers

Fig. 12 – Ariane (photo – Arianspace).

are trying to come up with specifications which will provide for the needs of satellite launches from the 1990s through into the next century.

In an attempt to reduce the costs of satellite launches, the North American Space Administration introduced its Space Transportation System in 1981. A re-usable launch vehicle, colloquially named the 'space shuttle', it can carry loads of around 1350 kg initially into a low circular orbit about 290 km above the earth, and then release them to travel on into a geostationary orbit. Shuttle flights carry both commercial cargo and passengers, frequently 'spacelab' scientists who are given the chance to carry out experimental projects in space. They are also used for military purposes. Although the launch vehicle which carries the satellites into the low 'parking' orbit is re-usable, the upper stages, known as Payload Assist Modules (PAMs) which carry the payloads on into geostationary orbit, are expendable. Several different PAMs are available and planned, which will allow the Space Transportation System to place payloads of up to 5500 kg into geostationary orbit. Although it seems obvious that a re-usable launch vehicle like the STS space shuttle should be much cheaper to use than expendable systems such as Ariane, there are in fact many parts of the STS that cannot be recovered, and much refurbishing of the shuttle has to be carried out after each flight. Current shuttle prices are well below those of Ariane ($14–20 million per satellite for a dual launch), but it is known that the US government is heavily subsidising STS, and costs for later shuttle launches are likely to increase significantly, whereas costs for Ariane are predicted to fall.

The Russians have made it known that they would like to get into the commercial satellite launching business, but so far they have had little success, in spite of quoting rates that undercut those of both European and American launch systems. As well as the problems of access for men and equipment to the heavily guarded soviet launch sites, there are engineering problems that make the Russian offers less desirable than they might at first seem. American and European launch organisations invariably use equatorial launch sites, so that when the satellite is launched in a west–easterly direction the surface velocity of the earth at the equator can be used to help the launch rocket to

Fig. 13 – Shuttle leaving launchpad (photo – NASA).

achieve the necessary escape velocity to free it from the effects of the earth's gravity, and to assist it to reach the velocity required to maintain it in orbit. The Russian launch sites are all in latitudes a long way north of the equator, so different launching techniques are required. As far as is known, extremely powerful rockets are used to lift the satellites directly into a near geostationary orbit, but the payloads that can be sent into geostationary orbit by this technique are believed to be smaller than is usual with the western systems.

Many private firms in America and Europe are currently considering ways of achieving lower cost launches for the large numbers of satellites that will be sent into orbit over the next few years. There are reports that fully re-usable systems are being planned which will take 4500 kg payloads into geostationary orbit at one twentieth of the cost of the shuttle. Whether these plans come to fruition remains to be seen.

2.3 PROPULSION METHODS

Rocket motors

All launch rockets need to turn some form of stored energy into kinetic energy. In the case of present-day rockets the energy is stored as chemical energy (e.g. hydrogen + oxygen) and is released as thermal energy which, in turn, is converted to kinetic energy in the motor exhaust nozzle. Overall conversion efficiencies of the order of 30%–40% are achieved.

The thrust F developed by the motor is then given by the equation:

$$F = V_e (\mathrm{d}M/\mathrm{d}t) \tag{2.1}$$

where V_e is exhaust velocity and $\mathrm{d}M/\mathrm{d}t$ is the rate of consumption of fuel. Note that, if $\mathrm{d}M/\mathrm{d}t$ is in kg/s, V_e in m/s, then F is in Newtons. The designer of the rocket therefore needs to arrange for as high a value of exhaust velocity as possible, commensurate with a given fuel consumption, in order to achieve the maximum thrust.

Although solid-fuel rockets have certain advantages for military applications, better performance at a given cost is available from liquid fuelled motors, and we shall, therefore, concentrate on the latter type.

Today's practical liquid-fuel rockets differ little in basic design from the designs suggested early this century. Fuel is injected into a combustion chamber where it is burned and where a high-pressure low-speed gas is transformed into a low-pressure gas moving at high speed. The means of achieving this is shown in Fig. 14.

With the symbols for gas conditions as defined above, by equating the energy within the combustion chamber to that in the exhaust the velocity of the exhaust gas V_e can be shown to be

$$v_e = \sqrt{[2\gamma/(\gamma - 1)]RT_c[1 - (P_e/P_c)^{(\gamma-1)}/\gamma]} \tag{2.2}$$

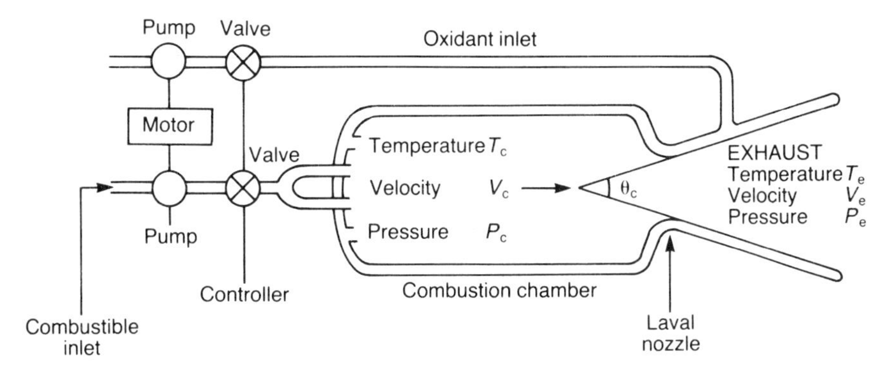

Fig. 14 – Basic rocket motor.

where γ is the ratio of the specific heat of the gas at constant pressure C_p to that at constant volume C_v, and R is the gas constant which equals $C_p - C_v$. Values for γ and R are more or less fixed for all fuels, and the designer needs to select fuels to give a high combustion temperature T_c and to design the exhaust nozzle to give maximum expansion ratio, thus minimising the term P_e/P_c. Typical figures for an oxygen/kerosine motor might be:

$$\gamma = 1.1$$

$$R = 370 \text{ J kg}^{-1}\text{K}^{-1}$$

$$V_e = 85.6\sqrt{T_c[1-(P_e/P_c)^{0.1}]}. \tag{2.3}$$

The maximum gas expansion ratio which can be achieved in practice is about 100:1, so that $P_e/P_c = 0.01$. $V_e = 85.6\sqrt{T_c(1-(0.01)^{-1})}$ and thus $V_e = 52\sqrt{T_c}$. For the fuel considered, $T_c = 3600$ K and $V_e = 3120$ m/s. In practice, various thermal/aerodynamic losses in the motor cause a loss equivalent to 10% in T_c, and the final exhaust velocity is about 2960 m/s.

A useful figure of merit for the performance of a rocket motor is its 'specific thrust', defined as S = motor thrust/fuel consumption per second. If motor thrust is in kg weight, and fuel consumption in kg/s, then specific thrust is in kg weights/kg. If motor thrust is expressed in newtons, specific thrust is in newtons/kg and is numerically equal to exhaust velocity in m/s. Specific thrust is therefore numerically equal to exhaust velocity divided by g.

For the above example, specific thrust S = 2960/9.81, i.e. about 300 kg wt. s/kg.

It is difficult to provide cooling for rocket motors operating at such high temperatures, but this is normally accomplished by feeding one of the fuel components through the walls of the motor. The oxidant is usually used as the coolant because a greater quality of oxidant is carried compared with combustible, but in kerosine or alcohol type motors, the combustible rather than the oxidant may be used.

A large expansion ratio is needed in order to obtain high efficiency, and the critical feature of the motor design to achieve this is the so-called Laval nozzle. The designer needs to determine the optimum compromise between a short pipe with relatively large angle θ_e (shown in Fig. 14) and a long pipe with a smaller flare angle. With a short, wide nozzle, power is lost owing to divergence of the jet efflux; whereas with a longer nozzle, power is lost through aerodynamic friction over the large surface area of the nozzle. The practical compromise for a simple nozzle with constant flare is θ_e within the range 20°–30°, but improved motors of recent design have nozzles with 'concave' flare.

Fig. 14 shows a diagramatic presentation of fuel injection with four jets. Practical motors are more complex, trying to achieve the best possible mixture of combustible and oxidant as quickly as possible. This requires many jets; for example, the V–2 motor had dozens of oxidant and combustible jets in each of its 18 precombustion chambers.

Rocket fuels

The fuels most commonly used in liquid fuelled motors are a hydrocarbon (kerosine or alcohol) or liquid hydrogen for the combustible, and liquid oxygen, nitric acid, or hydrogen peroxide for the oxidant; but there are many other possibilities.

Some alternative oxidants include fluoride or chlorine, which give higher performance than oxygen at the expense of more difficult engineering problems. Alternative combustibles include powdered metals (boron, aluminium), fed to the motor suspended in a liquid base, and hydrazine (N_2H_4) or dimethylhydrazine $(CH_3)_2N_2H_2$. The latter has been used for example in the 'Vanguard' and 'Jupiter–C' motors.

Characteristics of some fuels are given in Table 2.3. The corrosive/toxic/temperature range of the fuel combinations is clearly a source of difficulty in handling these materials, especially in the considerable quantities needed for launch vehicles. The ratio of masses shown, oxidant to combustible, is that required for complete combustion.

Hydrogen peroxide has been used as an oxidant, although it is rather unstable and is not normally employed in large motors. It does decompose (in the presence of a suitable catalyst), giving oxygen and steam at a temperature of about 400°C. Used as a mono-propellant in this way it is a suitable fuel for driving fuel pumps for large rocket motors. More recently, superior monopropellants have been developed, such as propyl nitrate and ethylene oxide.

For engines which require repeated start operation, hypergolic fuels are useful. These are auto-igniting and tend to have complex chemical structures. One such combustible is 'Tonka' which is composed of triethylamine and

Table 2.3 Some rocket fuel characteristics

Oxidant	Boiling point	Combustible	Boiling point	Ratio: wt oxidant / wt combustible	Combustion temperature K	γ	R J kg^{-1} k^{-1}	Specific thrust S kg wt. s/kg
Nitric acid	86	Kerosine	60–90	5.3	3 000	1.16	330	260
Liquid oxygen	−183	ethyl alcohol	78	2.1	3 300	1.12	370	290
Liquid oxygen	−183	kerosine	60–90	3.4	3 600	1.11	370	300
Liquid oxygen	−183	liquid hydrogen	−253	8.0	—	—	—	370
Liquid oxygen	−183	dimethyl hydrazine	63	—	3 500	1.15	390	290
Liquid fluorine	−182	dimethyl hydrazine	63	—	4 600	1.2	420	340
Liquid fluorine	−182	liquid hydrogen	−253	9.5	—	—	—	420

xylidine in equal parts. Such fuels produce power output similar to that of kerosine.

One of the most important design points in a rocket motor concerns the method of fuel ignition. Any delay in firing the mixture once fuel flow has started can result in an explosion in the combustion chamber which might easily cause complete destruction of the rocket on the launch pad. Correct starting requires ignition across the whole of the combustion chamber within about 20 ms of the nominal ignition time.

In practice, launch is accomplished using multi-stage rockets. That such rockets can achieve higher speeds than single-stage rockets is evident from the fact that energy is not expended in accelerating the casing of the first stage(s), the latter being jettisoned when fuel is expended.

Solid-fuelled motors have the advantage of simplicity and comparatively low cost. They may also be stored for long periods without special difficulties. In these respects they offer advantage over liquid-fuelled motors, especially when using liquid oxygen and/or hydrogen. On the other hand, they lack the power of liquid-fuelled motors and are unsuitable where a stop/restart capability is needed. The constructional arrangement commonly used is shown in Fig. 15.

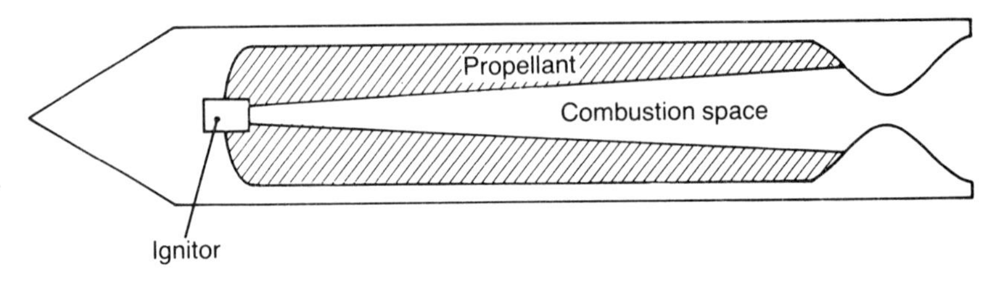

Fig. 15 – Construction of solid-fuelled rocket motor.

REFERENCES AND BIBLIOGRAPHY

[1] *IBA Technical Review* No. 11
[2] *IBA Technical Review* No. 18
[3] *EBU Technical Review* Aug. 1983
[4] *IBA Broadcast Engineering Notes* No. 2 (out of print)

3

Planning a satellite broadcasting service

Having established that for direct broadcasting from satellites there are many advantages to using a geostationary orbit, and having shown that it is in fact practicable to launch satellites into such orbits, we will now consider some of the other important factors which need to be thought of before such services can be introduced.

3.1 CHOICE OF FREQUENCY BAND

First let us think about the frequency bands which could be used. The Radio Regulations drawn up by the International Telecommunications Union (ITU) are the 'bible' so far as international frequency usage is concerned, and these regulations show that only a very limited number of bands have been allocated for broadcasting from space. The frequency bands are shown in Table 3.1.

For the purpose of regulation, the ITU divides the world into three regions, as shown in Fig. 16, and there are many instances where different regulations apply to different regions, which can make for particular difficulties when satellite broadcasting is being considered.

The first band shown in Table 3.1, 620–790 MHz, might at first seem to be the most suitable, since the engineering of UHF transmitters and receivers is well understood, and the economies of large-scale production have already

Table 3.1 ITU frequency bands for broadcasting from space

620–790 MHz, part of Band V
2500–2690 MHz, with some restrictions
11.7–12.5 GHz
22.5–23 GHz (Not region 1)
41–43 GHz
84–86 GHz

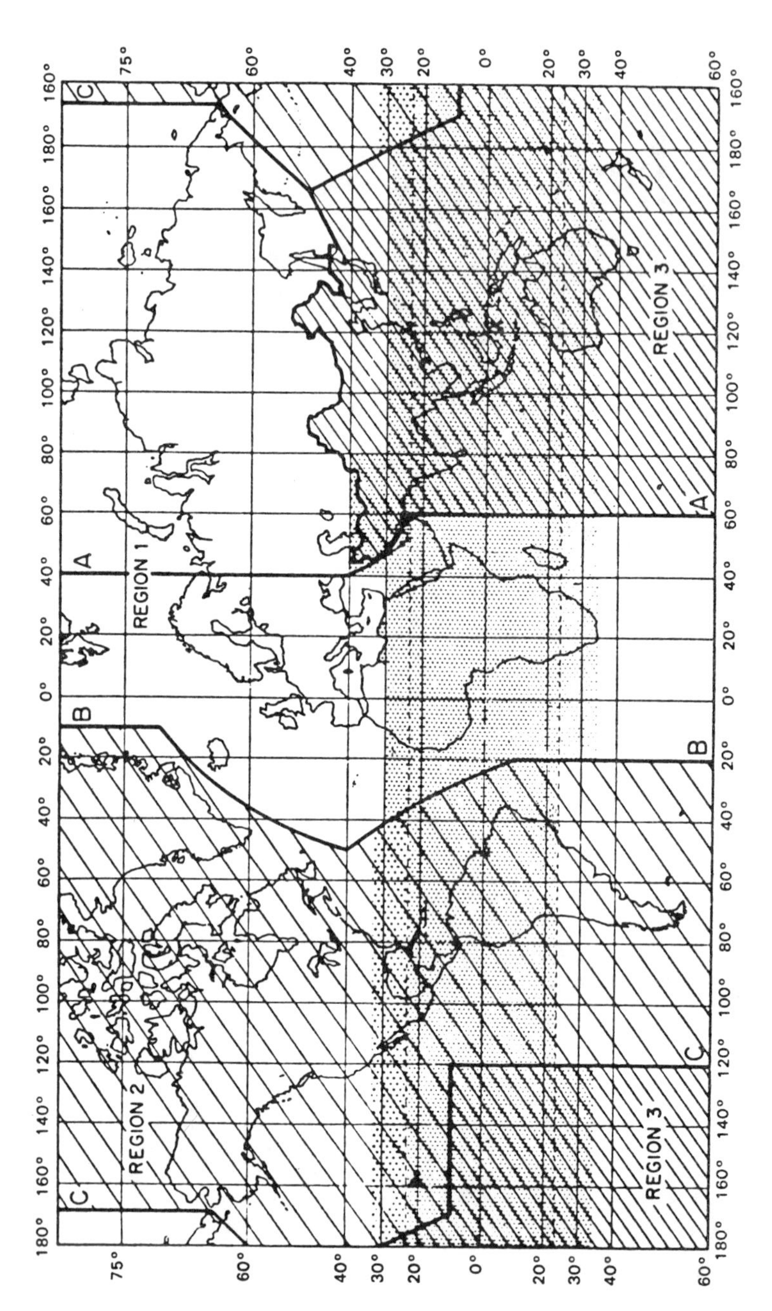

Fig. 16 – ITU region map.

made low-cost receiving equipment a reality. Indeed, the UHF band was used successfully for satellite broadcasts as long ago as 1975 when the Americans provided a satellite for the Indian Satellite Instructional Television Experiment (SITE). Frequencies in the UHF band, around about 860 MHz, were used to beam signals into some two and a half thousand villages scattered over six Indian states, where good reception was obtained with relatively primitive aerials, many of which were fashioned out of wire netting. Unfortunately, though, we are unable to consider using the UHF band, because broadcasters throughout Europe are already using these frequencies for terrestrial television broadcasts, and readers will realise the interference problems that might be caused to these existing services if satellites were to start beaming down relatively high-powered signals on the same frequencies.

The second frequency band in Table 3.1, 2500–2690 MHz, is hedged around with various restrictions, and is not all available in Region 1, so that we in Europe have to discount this band as a possibility for direct broadcasting from satellites.

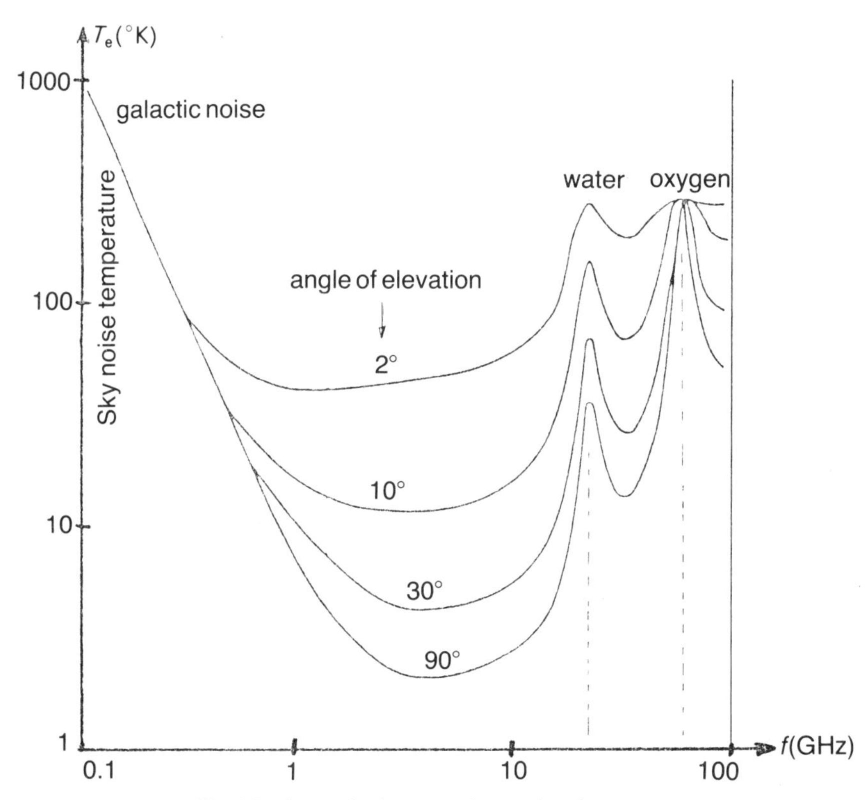

Fig. 17 – Atmospheric attenuation against frequency.

The 22.5–23 GHz band is not allocated for use in Europe, so cannot be considered for our purposes, but the Japanese are already seriously considering the use of this band for future high-definition television services.

It is expected that the 41–43 GHz and 84–86 GHz bands will one day provide many wideband satellite channels, in spite of the difficulties of propagation through the atmosphere that are encountered at such high frequencies (Fig. 17). At the present time, however, the technology does not exist for the manufacture of low-cost receiving equipment or high-power transmitting equipment for use at these frequencies. Even the most optimistic engineers are not predicting the widespread use of these frequencies until the next century.

An experimental satellite ATHOS (Application Technologique Hyperfrequences en Orbit Synchrone), due to be launched via Ariane in 1986, will include an experimental module which will permit propogation measurements in the 20, 40, and 90 GHz bands, and this should improve the state of our knowledge considerably.

Thus, by a process of elimination we are left with the band 11.7–12.5 GHz as the only possibility for satellite broadcasting in Europe at the present time, and to make the best possible use of this band a World Administrative Radio Conference (WARC'77) was held in 1977.

3.2 SPECTRUM PLAN

WARC'77 produced a complete plan for satellite broadcasting in Regions 1 and 3, using the 12 GHz band. In 1977 the Americans felt that it was too early to formulate plans for Region 2, and it was not until 1983 that a plan for satellite broadcasting in the Americas was agreed at the Regional Administrative Radio Conference in Geneva. Learning from European experience, the plan for Region 2 was deliberately made less detailed than the WARC'77 plan, and instead of allocating specific channels to each country, the plan allocates blocks of frequencies, which may be used in various ways. This method of allocation will make it easier for American broadcasters to introduce services requiring wider channel bandwidths than are currently available in Europe.

The WARC'77 plan divided the band between 11.7 and 12.5 GHz into 40 overlapping channels, and generally allocated five of these 27 MHz-wide channels (Fig. 18) to each country in Europe, permitting each country to broadcast up to five different television services (Tables 3.2, 3.3). It was assumed that each channel would be used to carry one frequency-modulated PAL or SECAM television signal together with its accompanying sound signal, but the planners took care not to preclude the use of other systems, provided that they cause no more interference to other users than that caused by the system stipulated in the plan. This decision turned out to be remarkedly

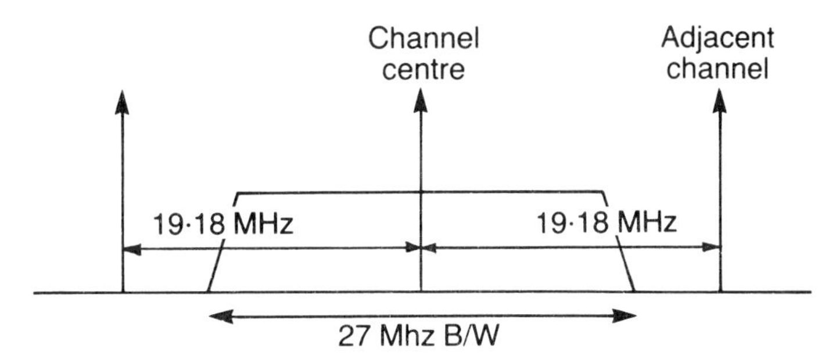

Fig. 18 – Spectrum plan.

Table 3.2 Correspondence between channel numbers and assigned frequencies for the 12 GHz satellite broadcasting band

Channel No.	Assigned frequency (MHz)	Channel No.	Assigned frequency (MHz)
1	11 727.48	21	12 111.08
2	11 746.66	22	12 130.26
3	11 765.84	23	12 149.44
4	11 785.02	24	12 168.62
5	11 804.20	25	12 187.80
6	11 823.38	26	12 206.98
7	11 842.56	27	12 226.16
8	11 861.74	28	12 245.34
9	11 880.92	29	12 264.52
10	11 900.10	30	12 283.70
11	11 919.28	31	12 302.88
12	11 938.46	32	12 322.06
13	11 957.64	33	12 341.24
14	11 976.82	34	12 360.42
15	11 996.00	35	12 379.60
16	12 015.18	36	12 398.78
17	12 034.36	37	12 417.96
18	12 053.54	38	12 437.14
19	12 072.72	39	12 456.32
20	12 091.90	40	12 475.50

Note: UK channels are 4, 8, 12, 16 and 20, orbit position 31°W, polarisation left hand circular.

far sighted, as most European countries eventually agreed to use a system that provided for multiple sound channels to accompany each television picture, and that they would use the Multiplexed Analogue Components (MAC) transmission system so as to give viewers the chance to receive higher quality pictures than the PAL or SECAM systems, used for terrestrial broadcasting, could provide. Table 3.4 summarises the plan.

Table 3.3 Orbital positions, channel assignments, and polarisations for countries of western and southern Europe

Orbital position	Lower half (11.7–12.1 GHz) Left-hand circular polarisation	Right-hand circular polarisation	Upper half (12.1–12.5 GHz) Left-hand circular polarisation	Right-hand circular polarisation
37° West	Andorra 4, 8, 12, 16, 20	San Marino 1, 5, 9, 13, 17		Monaco 21, 25, 29, 33, 37
		Liechtenstein 3, 7, 11, 15, 19		Vatican 23, 27, 31, 35, 39
31° West	Portugal 3, 7, 11, 15, 19	Ireland 2, 6, 10, 14, 18	Iceland 21, 25, 29, 33, 37	
		United Kingdom 4, 8, 12, 16, 20	Spain 23, 27, 31, 35, 39	
19° West	West Germany 2, 6, 10, 14, 18	France 1, 5, 9, 13, 17	Switzerland 22, 26, 30, 34, 38	Belgium 21, 25, 29, 33, 37
	Austria 4, 8, 12, 16, 20	Luxembourg 3, 7, 11, 15, 19	Italy 24, 28, 32, 36, 40	Netherlands 23, 27, 31, 35, 39
5° East	Finland 2, 6, 10	Turkey 1, 5, 9, 13, 17	Nordic 22, 24, 26, 28, 30, 32, 36, 40	Cyprus 21, 25, 29, 33, 37
	Norway 14, 18			
	Sweden 4, 8 Denmark 12, 16, 20	Greece 3, 7, 11, 15, 19	Sweden 34 Norway 38	Iceland 23, 27, 31, 35, 39

Table 3.4 Summary of WARC plan

Frequency band 11.7–12.5 GHz
40 overlapping channels – 27 MHz wide – 19.18 MHz centres
5 channels per country
Circular polarisation
31 dB co-channel interference protection
15 dB adjacent channel interference protection

As well as allocating frequencies to each country, the WARC specified the orbital position, the power, the transmitting aerial directivity pattern, and the polarisation of the signals from each satellite, and these characteristics were designed so as to minimise interference between satellites, whilst making the best possible use of the available bandwidth. In order to calculate the amounts of interference produced by any particular satellite at any orbital position the planners came up with a figure of merit known as the 'equivalent protection margin', which takes into account the effects of both co-channel and adjacent channel interference. The equivalent protection margin is officially defined as the value in dB by which the ratio of wanted signal to total interference exceeds the agreed co-channel protection ratio. The WARC planners decided that when calculating the equivalent protection margin for direct broadcast satellites they would use figures of 31 dB for co-channel and 15 dB for adjacent channel interference protection ratios, and it is generally agreed that these figures represent a very high standard of protection.

Although the bandwidth of each channel is 27 MHz, it was found possible to place the centre frequencies of these channels only 19.18 MHz apart whilst still maintaining the required interference protection ratios. This overlapping allows 40 channels to be assigned, and leaves a guard band at each end of the satellite band to provide protection to and from other services.

3.3 ORBITAL SLOTS

If viewers are to be able to receive interference-free pictures with small receiving aerials the satellites must be spaced sufficiently far apart for signals from an adjacent satellite to have no appreciable effect on reception of the wanted signal. On the assumption that viewers would use aerials equivalent in performance to a 0.9 metre diameter parabolic dish (beamwidth about 2°), and that the minimum practicable transmitting beam would be circular, with a half power beamwidth of 0.6°, the WARC planners decided that broadcasting satellites should be positioned at 6° intervals around the orbit. Some of the orbital positions allocated to European and north African countries are shown in Table 3.5 and Fig. 19.

Notice that Albania and Yugoslavia are the only European countries

Table 3.5 Orbital positions

Position	Countries
5° E	Norway, Sweden, Finland, Denmark
1° W	Poland, East Germany, Czechoslovakia, Hungary, Roumania, Bulgaria
7° W	Albania, Yugoslavia
19° W	France, West Germany, Belgium, Netherlands, Luxembourg, Italy, Austria, Switzerland
25° W	Libya, Tunisia, Algeria, Morocco
31° W	United Kingdom, Ireland, Spain, Portugal, Iceland

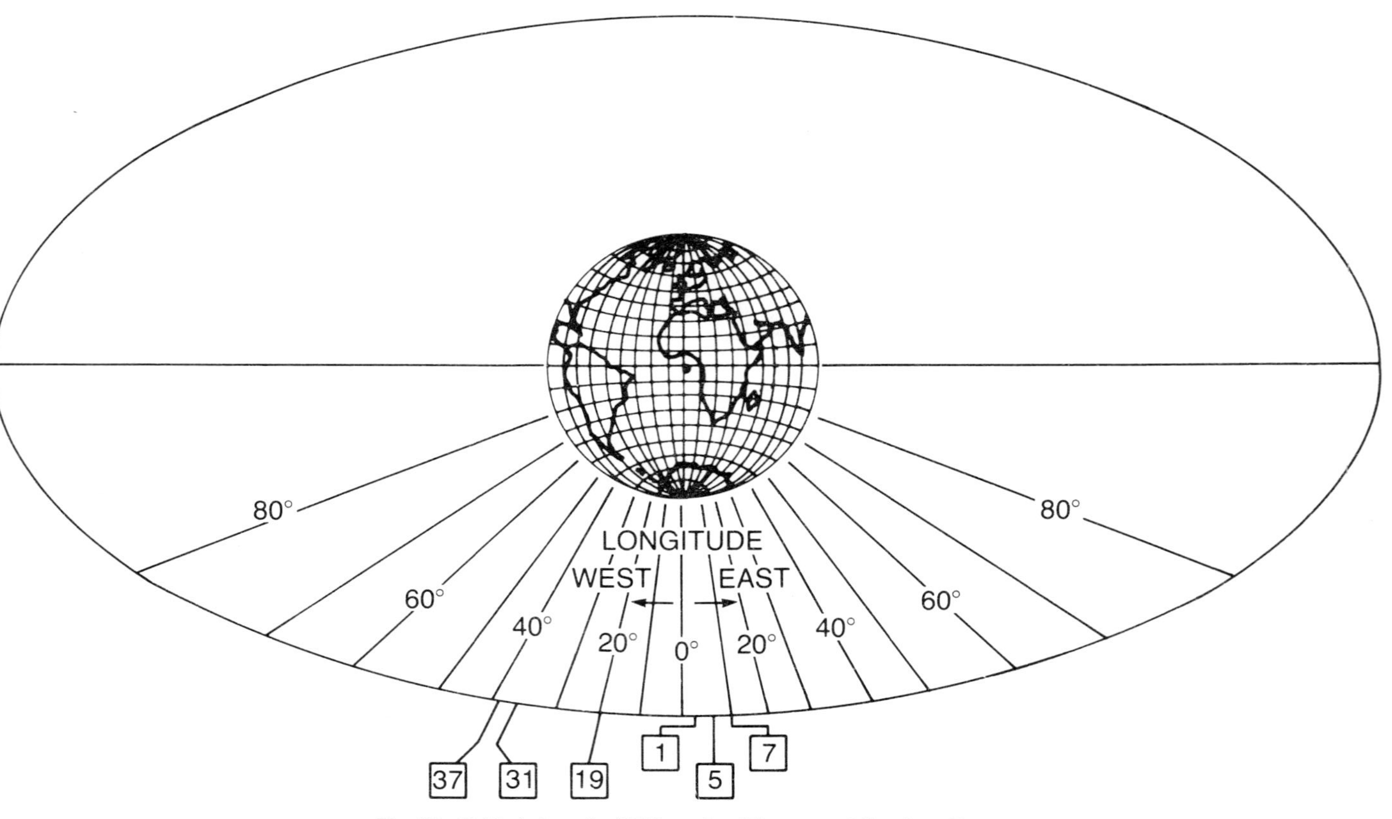

Fig. 19 – Orbital chart for DBS serving Western and Southern Europe.

with satellite allocations in the 18° spacing between 1° W and 19° W. This came about because it proved impossible, using 6° spacing, to fit in all the required satellites for the many countries of Central Europe whilst maintaining the desired equivalent co-channel protection ratio of 31 dB. The 7° W and 13° W slots are mainly used for African states.

3.4 COVERAGE AREAS

In general the WARC planners tried to arrange that the satellite beams covered, as far as was technically possible, only the countries for which the transmissions were intended. This was considered particularly important since some of the east European countries made it quite plain that they did not want other countries beaming in programmes intended for their residents. The narrowest possible beamwidth for a 12 GHz aerial on a satellite that will provide coverage of the whole of the United Kingdom, and that can be realistically achieved at the present time, is about ± 0.3°. Fig. 20 shows that this provides an illuminated area of 'footprint' that is adequate to cover the whole of the country. The power flux density at the edge of the service area is generally assumed to be half that at the centre of the beam, so giving rise to the expression 'half-power beamwidth'. The beamwidth that can be obtained is a function of the size of the aerial that can be carried in the launch vehicle and mounted on the satellite. Present day launchers can carry aerials of around 3 m diameter, giving a minimum beamwidth of about ± 0.3°. It may prove possible in the future to build bigger aerials in space, so that satellite beams may be narrower, making it possible to serve smaller regions than can at present be envisaged. Some allowance has to be made for the fact that even the best-regulated satellite will not be able to keep exactly on its allotted station, and although direct broadcast satellites should be much better than some of their predecessors at station-keeping, it is still considered prudent to allow for a margin of about 0.1° when attempting to plan the coverage of any particular country. Before a footprint can be drawn it is necessary to make certain assumptions about the performance of the receiving equipment and about the power flux density that the satellite will be able to provide at the earth's surface. The WARC planners assumed that domestic receiving installations would have a certain level of performance, or figure of merit, and the footprints in Fig. 20 were calculated on the basis of these assumptions. Over the years that have passed since 1977 the design of receivers and the availability of suitable microwave semiconductor devices have improved, so that we expect domestic receivers to have much better performance levels than were planned for. This improvement in receiver technology effectively means that viewers will be able to use smaller dishes or to receive satellite signals well outside the area specified in the WARC plan. The footprint shown in Fig. 21 represents an up-to-date assessment of what should be possible with modern domestic receiving equipment using the currently planned UK satellite.

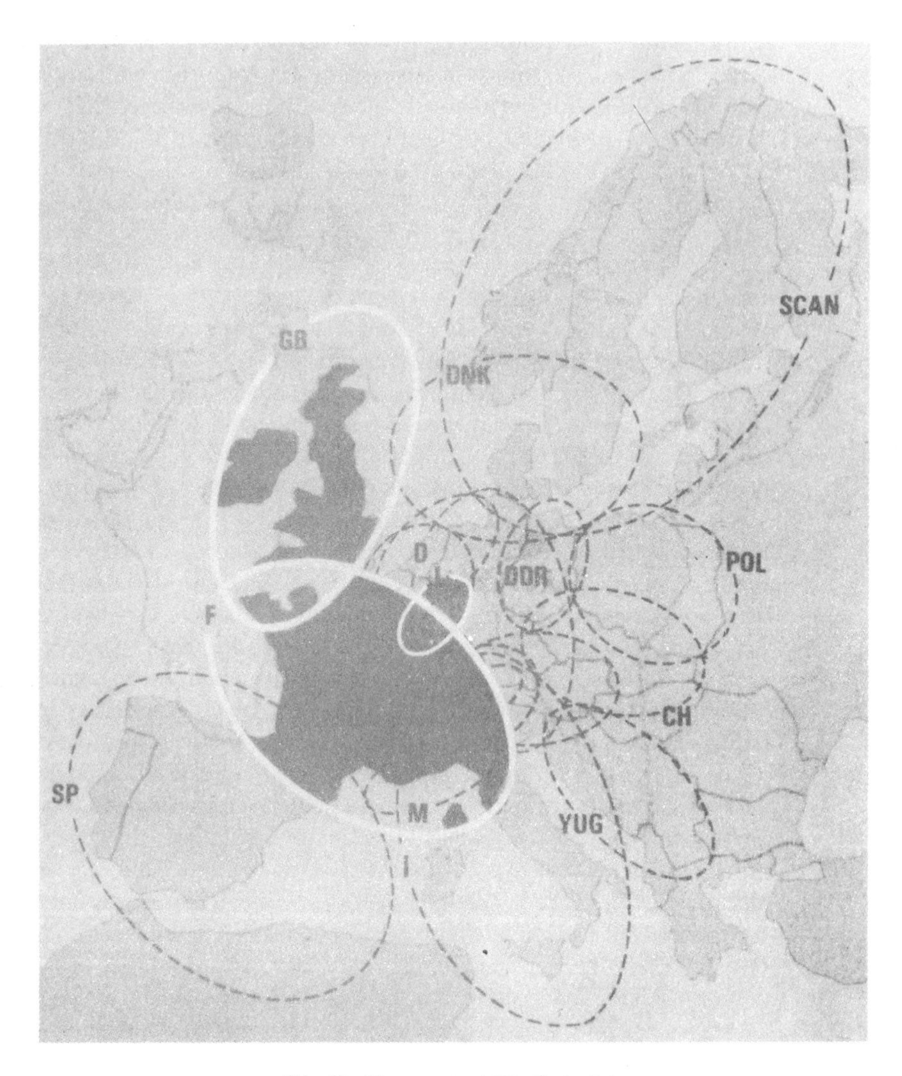

Fig. 20 – European satellite footprints.

3.5 RECEIVING EQUIPMENT

A useful figure of merit for receiving equipment that is frequently used in this context is the G/T or gain/noise-temperature ratio. This is essentially a comparative measure of the effectiveness of the complete receiving installation, and is defined as the ratio of the gain of the antenna in dB (relative to an isotropic source) to the effective noise temperature of the receiver. G/T ratios are therefore measured in units of dB/K, i.e. decibels per (degree) Kelvin. For

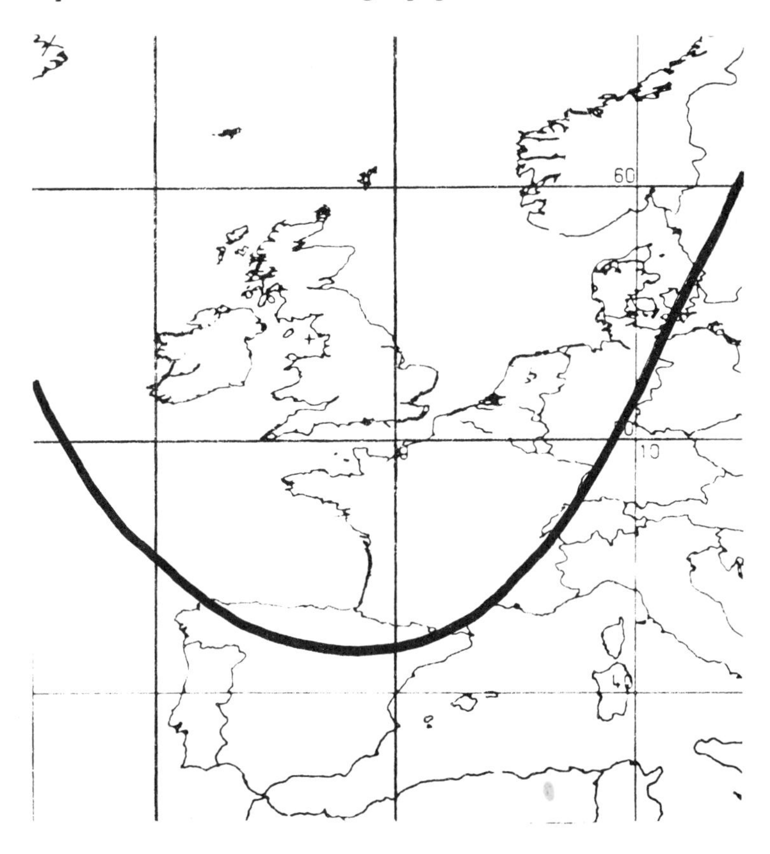

Fig. 21 – Approximate boundary of UK satellite reception using 0.9 metre dish and modern receiver front end. (C/N ratio ~ 12dB).

any given installation of a particular G/T ratio you can choose to have a large dish with a relatively poor amplifier, or a smaller dish and a higher-performance low-noise front-end. We shall look at the measurement of noise in detail in a later chapter, but for the time being it is sufficient to note that the initial 1977 WARC plans were formulated on the assumption that domestic receivers would achieve G/T ratios of 6 dB/K. In reality it is likely that developments that have taken place since 1977, especially in the field of low-noise R.F. amplifiers, will mean that by the time satellite broadcasting eventually starts, domestic receiving equipment will be able to achieve G/T ratios of around 10 dB. The footprints shown in Figs 20 and 21 were drawn on the assumption of G/T ratios of 6 dB/K and 10 dB/K respectively, and the increase in coverage area provided by the later equipment is substantial. Many private viewers within the service area may choose instead to take advantage of the improvement in receiver front-end performance by using a smaller receiving dish than the 0.9 metre receiving dish that the planners anticipated. Smaller aerials are likely to be more convenient to install and more acceptable environmentally than their larger brothers, so this may well be an important factor in determining the acceptability of satellite broadcasting to the general public.

Typical minimum G/T figures for reception of low-power communication satellites are 25–40 dB/K, and reception from distribution satellites with power of a few tens of watts requires G/T ratios of around 10–25 dB/K. A domestic installation with a 0.9 metre dish and a GaAsFET low-noise amplifier should provide a G/T ratio of 10 dB; this will leave a good margin for poor reception conditions, but many viewers in the UK may choose to use poorer equipment with smaller dishes that will be less obtrusive but will have less 'in hand' to cope with poor reception conditions that may occur during periods of heavy rain or snowfall.

3.6 SATELLITE POWER

For planning purposes the 1979 WARC assumed that DBS satellites would produce a power flux density (pfd) of $-103\,\mathrm{dBW/m^2}$ at the edge of their service areas, with the pfd at the centre of the service area assumed to be 3 dB greater than at the edges. To clarify the term 'power-flux density', we can say that if an aerial of area 1 square metre were to be pointed directly towards the satellite it would intercept 103 dB less than 1 watt of power, which works out at around 50 picowatts. It is instructive to look at a typical DBS satellite system power budget, since this enables us to get a good idea of the sort of powers that are required at each stage of the system.

Let us start by asking what quality of picture we expect to get from our receiver, and we will then work 'backwards' to determine the satellite power that is needed to provide this sort of picture.

Pictures are graded technically on a five-point scale, as specified in Table 3.6.

If we say that pictures at grade $4\frac{1}{2}$ would be acceptable, this corresponds to a video signal-to-noise ratio of around 40 dB.

Table 3.6

Five-grade scale	
Quality	**Impairment**
5 Excellent	5 Imperceptible
4 Good	4 Perceptible, but not annoying
3 Fair	3 Slightly annoying
2 Poor	2 Annoying
1 Bad	1 Very annoying

For an FM system, assuming a receiver with a noise factor of 7 dB, and a WARC bandwidth of 27 MHz, a signal power input of around −104 dBW is required to provide pictures with a video signal-to-noise ratio of 40 dB. If an allowance of 0.5 dB is made for the inevitable losses in the aerial to receiver connection, this means that we are looking for a signal power of −103.5 dBW from the aerial.

A 0.9 metre diameter circular dish of typical 50–60% efficiency will have an effective area of around 0.3 m^2, so we must arrange that there is sufficient power flux falling on an area of 0.3 m^2 to give the required aerial output power of −103.5 dBW. In practice it will be prudent to allow for a small amount of loss, say 0.5 dB, to take account of any slight inaccuracies in aerial positioning, so we end up by needing a power flux density of around −98 dBW/m^2 at the aerial. The satellite must radiate sufficient power to ensure that this pfd is achieved at the earth's surface (i.e. the receiving aerial) after the satellite's signals have travelled a distance of around 39 000 km between satellite and receiving aerial.

If the satellite is considered, for the purposes of calculation only, to be an isotropic source of *P* watts at a distance *d* from the receiving aerial, then the pfd at the receiving aerial can be calculated from

$$pfd = \frac{p}{4\pi d^2}\ \mathrm{W/m^2} \quad \text{or} \quad p = 4\pi d^2 \times pfd,$$

which in our example works out at just over 3 MW. This calculation is merely the 'free-space' path loss, caused by the spreading out of the beam as it travels away from the satellite, and the 12 GHz signals will actually be attenuated by their passage through the atmosphere. This attenuation will vary considerably, depending upon whether skies are clear or cloudy and whether or not heavy rain or snow is falling; but observations have shown that the mean 'excess path loss' – as this atmosphere attenuation is called – is around 0.5 dB. We must therefore increase the power of our satellite to compensate for this, so that we effectively need about 3.5 MW of power to be radiated from the satellite. This power is only needed in the direction of the earth, and it therefore makes good sense to make use of a highly-directional aerial in order to achieve the extremely high effective radiated power of over 3 MW. A term frequently used in this context is Effective Isotropically Radiated Power, (EIRP), which is the product of the power supplied to an antenna and its gain in a given direction. The more gain that can be achieved from the on-board transmitting aerial, the better, since this will enable the lowest possible transmitter power to be used, and this is a very important factor since all our power will be coming from solar panels which have a fairly restricted area because they have to be fitted into the nose cone of the launch-rocket or into the 'cargo-bay' of the space shuttle. At the present time this means that the solar panels, hinged and folded with the greatest mechanical precision, have to fit into a 3-metre space when the satellite is being carried aboard Ariane, and into a 5-metre space when the shuttle is used.

The type of transmitting aerial that provides the most gain for the least space at frequencies of around 12 GHz is generally agreed to be the parabolic dish, and most satellites use one or more dish aerials to achieve the gain that they need to beam their microwave signals down towards particular regions of the earth's surface. The maximum size of the dish that can be used is again limited, and a typical practical design would be able to provide a typical beamwidth of 0.5°, giving a gain of about 45 dB at 12 GHz.

Once again we would be wise to allow for coupling losses between the transmitter and the aerial dish, and a figure of about 0.5 dB would be realistic. Taking this into account with the gain of our aerial we can calculate that for an EIRP of 3.5–4 MW we need an RF transmitter power of between 150 and 200 watts. Typical of this breed of direct-broadcast satellite is the Unisat design which British Aerospace are planning to build. Its initial design specification provides for two television transponders of 200 watts each plus six low-powered telecommunications transponders, but this specification may well be changed before UNISAT actually reaches the launchpad.

The most common device that is currently used for high-power microwave amplifiers is the thermionic travelling wave tube, and typical efficiencies that can be expected when these are suitably downrated to improve their longevity, are of the order of 0.3, often expressed as 30%. This means that to provide our satellite with 200 watts of transmitted RF power we will need at least 600 watts of electrical power from the satellite's solar cells. It can be seen that satellites carrying multiple transponders will make heavy demands on their power supplies, and the large-platform L-Sats that are being planned for later in the decade are expected to be able to provide around 7.5 kW from their solar panels, so as to be able to accommodate the demands of five high-power DBS transmitters as well as the power required for the necessary telecontrol and telemetry services. At the present time solar panels of a size that can fit into current launch vehicles cannot be expected to provide more than around 4.5 kW, but techniques of in-space assembly may enable much larger banks of solar panels to be used in the future. The output from solar cells falls off considerably over a period, so that it is necessary to provide far more power at the start of the satellite's life than might at first be thought necessary, to ensure that the output from the solar cells will be sufficient towards the end of its design lifetime, which may well be as long as seven years.

Although there is fairly commonly battery-backup for the low-power telecommunications satellites it is not practicable to use batteries for the relatively high-powered DBS equipment. Chemical-powered fuel cells are unsuitable because the chemicals would need to be replaced many times during the seven years of the satellite's life, and nuclear-powered generators are these days considered unsuitable for DBS on environmental grounds.

One significant problem with solar-powered satellites is that all power is lost if the solar panels are shielded from the sun for any reason. Since the satellite is orbiting the earth, and the earth is orbiting the sun, there will be times when the earth gets between the satellite and the sun, effectively putting the solar panels in shadow. These eclipses occur at around the local midnight at the longitude of the satellite, and only at the times of the spring and the autumn equinoxes when the sun, the earth, and the satellite are all in line. They can last for periods of up to 72 minutes at a time during a period of about forty days in each season. Fig. 22 shows how the duration of these eclipses varies day-by-day over the period of the equinox.

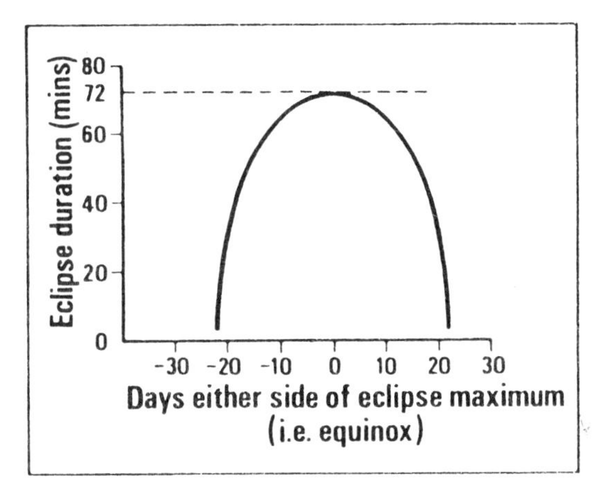

Fig. 22.

Since it would obviously be inconvenient if a direct-broadcasting satellite were to lose its power during the hours of television broadcasting, and since the 'late-night movie' is a popular feature of many nations' television services, some means had to be found of overcoming the eclipse difficulties, but fortunately for the satellite planners, the effect of the eclipse can be delayed if the satellite is placed to the west of its coverage area. Every degree by which the satellite is moved westwards causes the eclipse to be delayed by about four minutes, so if we are to be sure that the eclipse will never occur before midnight we need to move the satellite far enough west to give a delay of half of the 72 minute maximum eclipse period, which works out at around 9°, although because of variations in the length of the solar day it is better to allow for 11°. The WARC planners arranged for the UK satellites to be at 31° west of Greenwich, which makes any eclipse unlikely before about half-past one in the morning.

3.7 INTERFERENCE PROTECTION

To make the most effective use of the 40 frequency channels that are available in the 12 GHz band, a considerable amount of frequency re-use will be required, and to maintain the desired levels of interference protection against satellites in nearby orbital slots using similar frequencies, it proved necessary to use polarisation discrimination on transmission and receiving aerials. Experiments with OTS showed that each of two separate orthogonally polarised co-channel 12 GHz television signals could be received without noticeable interference from the other (i.e. a co-channel protection ratio of 30 dB) when appropriately polarised receiving aerials were used, on a clear day. As soon as any precipitation, such as rain, snow, or hail occurs, the cross-polar protection ratio could fall to less than 20 dB.

Circular polarisation was chosen for direct broadcasting from satellites rather than linear, because protection against signals of the opposite circular polarisation is maintained at angles which are well away from the main axis. If the satellite's beam rotates about its axis, this will have little or no effect on circularly polarised receiving aerials, whereas signals would deteriorate if orthogonally polarised aerial systems were used.

To maximise the interference protection provided by polarisation discrimination the WARC plan arranged to use different directions of polarisation on odd and even numbered channels throughout Europe, and to reverse these opposite polarisations for African coverage. In addition, it was arranged that satellites in adjacent orbit locations would use different polarisations, so that over European or over African coverage areas the same channel with the same polarisation is used only from alternate orbital slots, never from adjacent satellites.

Energy dispersal

The energy spectrum of the unmodified satellite transmission signal has a fairly broad frequency spectrum with several 'peaks', so a special low-frequency energy-dispersal signal is added to all transmissions to effectively flatten the peaks, spreading the energy more evenly over the spectrum. This prevents significant energy peaks being radiated at particular frequencies, making interference with other 12 GHz users less likely. A typical dispersal waveform might consist of a frame synchronous triangular waveform of 25 Hz with a deviation of 600 KHz p-p after modulation.

REFERENCES AND BIBLIOGRAPHY

[1] *IBA Technical Review* No. 11.
[2] *IBA Technical Review* No. 18.
[3] *EBU Technical Review* Aug. 1983.
[4] *IBA Broadcast Engineering Notes* No. 2 (out of print).

[5] *Final acts of the World Broadcasting Satellite Administrative Radio Conference, Geneva* 1977. I.T.U. Geneva 1977.

[6] *Analysis of* 1977 *Geneva Plan for Satellite Broadcasting at* 12 *GHz* Tech 3222-F/E. EBU Technical Centre, Brussels, April 1977.

[7] C.C.I.R. Recommendation 500.2. *Method for subjective assessment of quality of Television Pictures.*

4

The broadcast service

4.1 CHOICE OF A MODULATION METHOD

Terrestrial television broadcast services have traditionally used amplitude modulation, but the power resources available to direct broadcast satellites are likely to be extremely limited, and this means that it would be very difficult to provide enough AM transmitted power to give a satisfactory signal-to-noise ratio with the low-cost receiving installations around which DBS systems were planned. Using frequency-modulated signals the required signal-to-noise ratio can be achieved with some 20 dB less transmitter power than would be needed for AM signals. This power advantage turns out to be the crucial factor in deciding that FM signals shall be used for DBS, but there are also other significant gains to be made by the use of FM.

Although it is usually agreed that FM signals occupy a wider bandwidth than would the corresponding AM signals, it has been found that the protection ratio required against co-channel FM signals is some 20 dB less than that required for AM. To achieve television pictures which are free from interference in the presence of a co-channel signal generally requires the wanted signal to be around 45–50 dB greater than the potentially interfering AM signal. For frequency-modulated signals the same degree of protection can be achieved when the wanted signal is only about 30 dB greater than the interfering co-channel FM signal. Using FM therefore allows more efficient use to be made of the RF spectrum, since some degree of overlap can be tolerated, and, as Fig. 23 shows, the WARC plan arranged for FM signals with an overall bandwidth of 27 MHz to be transmitted at 19.18 MHz intervals. The well-known FM 'capture-effect' will ensure that receivers have no difficulty in remaining tuned to the centre frequency of the band.

It might have been thought sensible, in these days when digital techniques are so far advanced, to take the opportunity of the start of satellite broadcasting to introduce digital transmissions. The international digital studio standard (CCIR Rec 601) for television uses separate component signals, sampling luminance signals at 13.5 MHz and each of two colour difference signals at half this rate. With 8 bits per sample the bit rate required

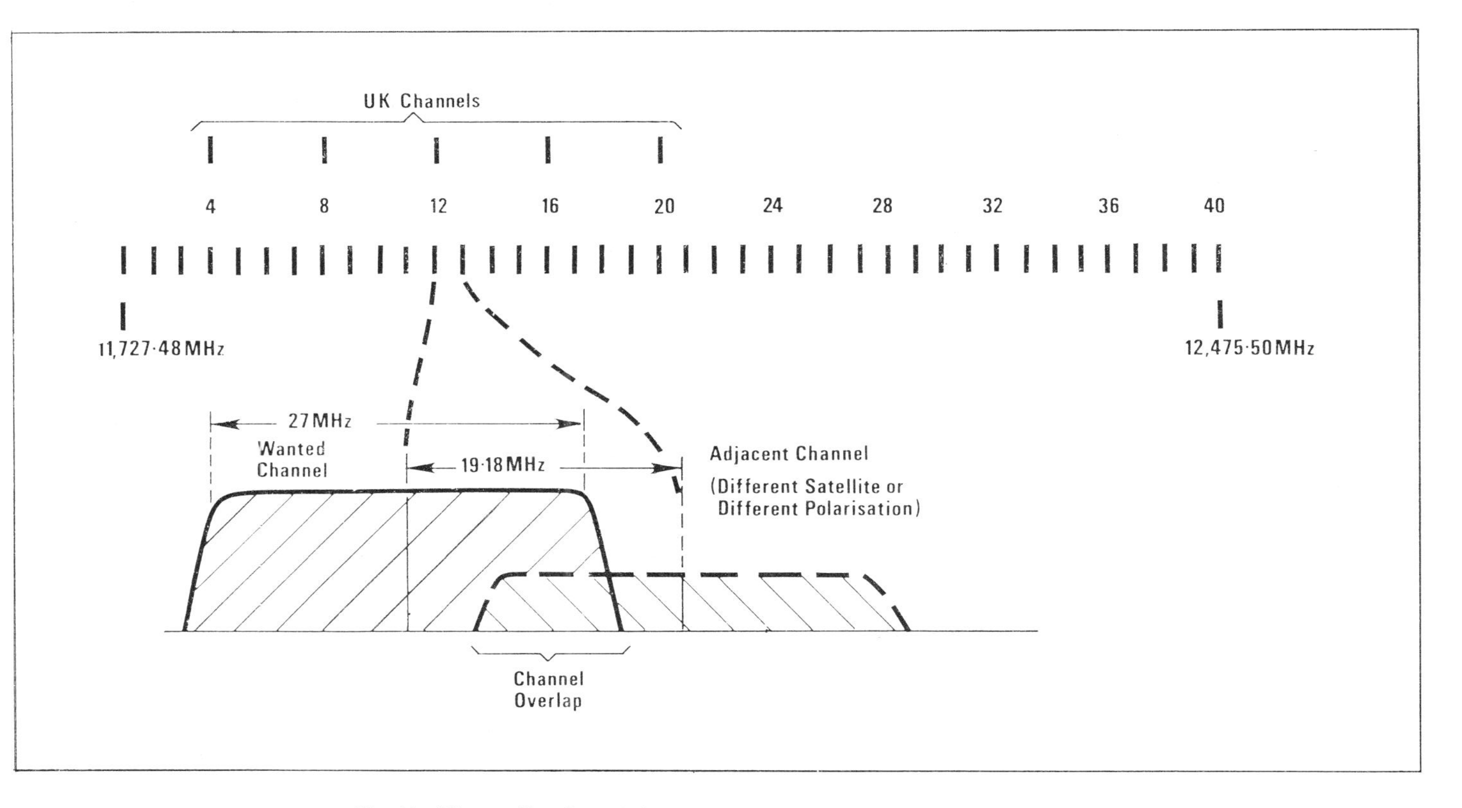

Fig. 23 – The satellite channels in the 12 GHz band as defined by WARC 1977.

for transmission is therefore $8 \times (13.5 + 6.75 + 6.75) = 216$ Mbit/s, which implies an RF bandwidth of about 100 MHz if such a signal is to be transmitted. The 27 MHz wide WARC satellite channels are obviously insufficient to carry these studio standard digital signals, but much work is being carried out in research laboratories throughout the world to try to achieve a reduction in the bit-rate of transmitted digital pictures. IBA engineers demonstrated the feasibility of transmitting 140 Mbit/s pictures with absolutely no visible degradation in 1981, and the BBC have since gone on to experiment with the use of 140 Mbit/s systems as part of their terrestrial distribution network. Such pictures still need an RF bandwidth of around 70 MHz, still too great for normal satellite transmission, although some people have been looking into the possibility of using two satellite channels to carry higher definition pictures at this sort of bit-rate. Experimental transmissions have been made with data rates as low as 34 Mbit/s, but so far the picture quality that is achievable leaves a good deal to be desired when compared with the quality of a normal broadcast signal. Although further improvements in bit-rate reduction will certainly take place over the next few years, it was not felt to be practicable to use digital transmission of video signals for the start of satellite broadcasting, and the decision was therefore made to use frequency modulation of analogue video signals.

4.2 THE MULTIPLEXED ANALOGUE COMPONENTS SYSTEM

The NTSC, PAL, and SECAM signals which are used for terrestrial television services suffer from some noticeable technical disadvantages, and some of these are emphasised when the signals are passed over a frequency-modulated satellite channel. Engineers from the UK's Independent Broadcasting Authority suggested that the start of direct-broadcast satellite services might provide a suitable opportunity for the introduction of a higher quality television system, which would be designed specifically to be free from the problems affecting the existing systems. It is always difficult to introduce changes to broadcast signals, since so many existing receivers are affected, but since the reception of satellite broadcasts will require some sort of adaptor to convert the 12 GHz FM signals into a form that can be accepted by conventional AM UHF/VHF receivers, this same adaptor could be designed to accept an improved signal which would provide higher quality pictures with suitable receivers.

The 1977 WARC plan did in fact make provision for the adoption of new broadcasting systems. Although WARC calculations were carried out on the assumption that 'standard' PAL/SECAM signals would be radiated, the final acts stated that the use of other systems was not precluded, provided that their use caused no more interference to other users than that permitted by the specified plan. To enable the effects of co- and adjacent channel interference to be assessed, the WARC plan provided the interference template shown in Fig. 24. At its simplest, any system which keeps its potentially interfering signals within the boundaries of the template should be acceptable.

The main problems that the present-day PAL and SECAM transmission systems present are cross-colour, cross-luminance, and chrominance noise.

Cross-colour occurs in areas of the picture containing fine detail, and is the generation of spurious coloured patterns caused by the inability of the receiver's decoder to separate colour information from the high-frequency black and white information. Fig. 25 shows how the colour information is interleaved in the frequency domain with the black and white information in a PAL signal.

Fig. 26 shows why it is very difficult for any receiver to separate the embedded colour information from the highly detailed black and white information in the vicinity of the colour subcarrier.

Another similar effect, cross-luminance, is caused by the residual colour subcarrier causing brightness variations where sharp colour transitions occur. This shows itself in the form of a dot-patterning effect along the edges where rapid changes of colour take place, especially noticeable when contrasting colour areas with vertical edges (e.g. vertical colour bars) are transmitted.

As explained earlier, all currently planned satellite broadcasting systems will use frequency modulation, because of the signal/noise ratio advantages which FM gives when transmitter powers are restricted, as they must be on satellites.

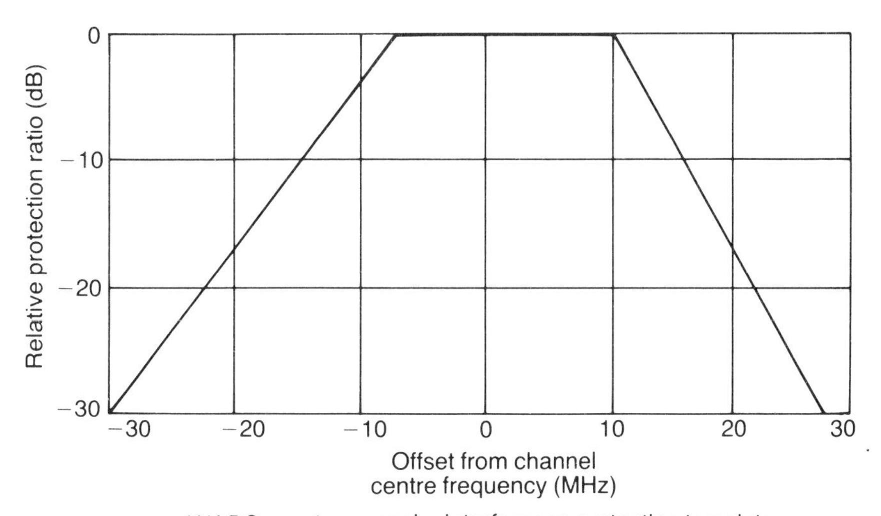

Fig. 24 – WARC spectrum mask – interference protection template.

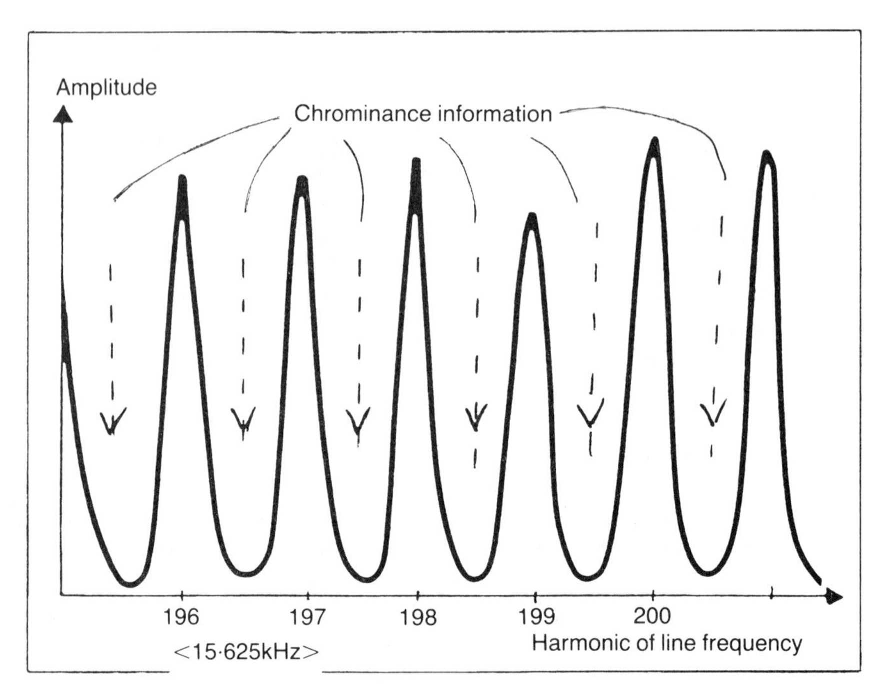

Fig. 25 – Luminance spectrum of 625-line PAL television signal (static picture) showing how colour information is inserted into gaps.

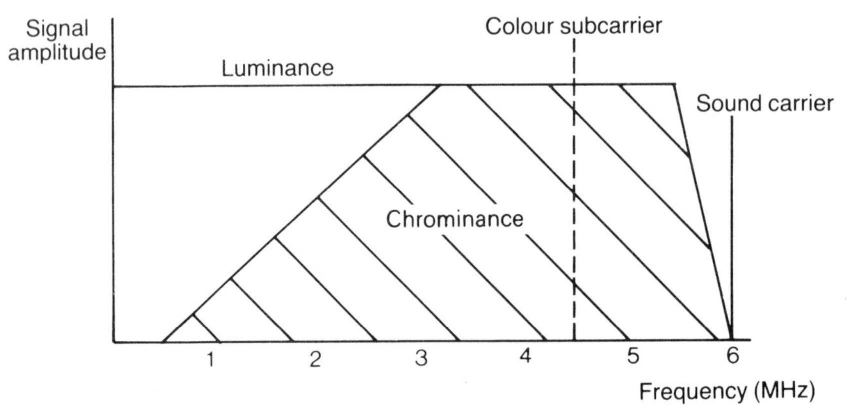

Fig. 26 – Baseband spectrum of PAL system I showing how colour information is mixed with luminance information, making it difficult for a receiver to separate colour from detailed black and white information.

It is well known that FM transmissions have a triangular noise spectrum, as shown in Fig. 27, and this implies that any system noise will be most apparent at the high-frequency end of the spectrum. Since PAL signals carry their colour information in the high-frequency parts of the picture spectrum,

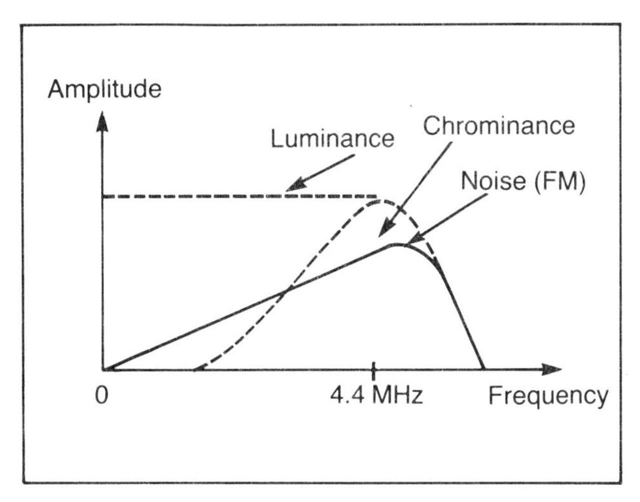

Fig. 27 – Noise spectrum of an FM signal, showing why noise will be most apparent in highly saturated coloured areas of a PAL picture.

PAL signals carried in FM channels are subject to chroma noise, which is particularly noticeable in large areas of saturated colour.

To overcome these problems, the Independent Broadcasting Authority developed, during the years 1981–84, a new system of transmitting television pictures, the Multiplexed Analogue Components (MAC) system, which works in a completely different manner to the existing PAL and Secam systems.

The term 'multiplexed analogue components' strictly describes just the vision signal, consisting of the analogue chrominance components (the colour information) and the analogue luminance component (the black and white information) of the picture. The off-screen photograph (courtesy of IBA) reproduced on the back cover of this book shows how the MAC system transmits the luminance and the chrominance signals during separate periods (Fig. 28), a technique known as 'time-division multiplex'. Since the two signals

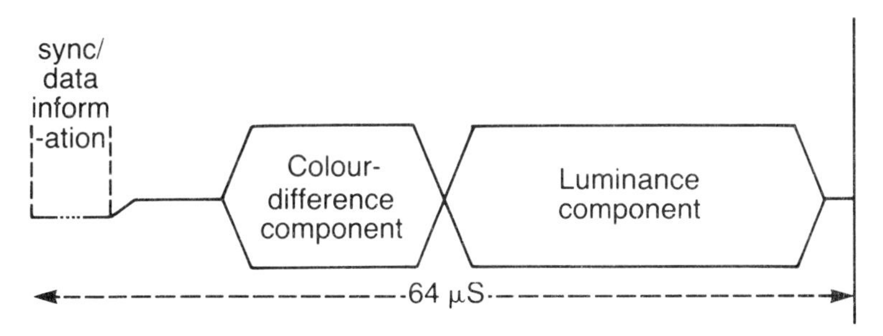

Fig. 28 – One line of MAC signal, showing how luminance and chrominance are transmitted separately.

are never transmitted together, there is no chance of interaction between the chrominance and the luminance parts of the picture, so cross-colour and cross-luminance cannot occur.

To enable the separate colour and luminance signals to fit into the standard 64 μs length of a television line, which is essential if compatibility with existing 625-line television receivers is to be maintained, both signals are time-compressed on transmission so as to pack them into less than 64 μs, and once they reach the receiver they are then expanded so that both the colour and the black and white parts of the picture once again fill a complete active television line. The major problem inherent in the technique of time-compression of the analogue signals is that as they are compressed in time the information which they contain has to be carried within a shorter period, so that the bandwidth that is needed to transmit these signals increases. This in itself is basically undesirable, since it is always a tenet of the broadcasting engineer's faith to keep the bandwidth requirements of any signal to a minimum, and an additional snag is that the wider the bandwidth that signals take up, the more susceptible to noise they become. The IBA work showed, however, that by choosing suitable compression ratios for each component signal it was possible to obtain first-class pictures with signal-to-noise ratios generally better than those possible for PAL or SECAM systems, without infringing the bandwidth limitations that had been laid down for satellite transmission.

The time compression (and later expansion) is achieved by sampling the analogue signal, storing the samples, and reading them out from the store at the appropriate time at a higher or lower clock frequency. Optimum results were obtained with the luminance signals time-compressed in the ratio 3 : 2, and with the chrominance components time-compressed in the ratio 3 : 1. The luminance component of each picture line is transmitted on its own line, together with just one of the two chrominance (colour-difference) component signals that normally make up a television picture line. In the MAC receiver's vision decoder the luminance and colour-difference signals are expanded, delayed as appropriate, and added together so that they are read out to give a picture which has luminance and 2 colour difference component signals on each line. The human eye is not particularly critical of coloured detail in a television picture provided that the black and white image is sharp, and critical viewing tests have shown that very satisfactory coloured pictures can be obtained with this technique of sending only one colour difference component on each television line. The big advantage of this method is that the time compression ratio does not need to be as great as it would have had to be if luminance and *both* colour-difference components were transmitted on every line, thus keeping bandwidth requirements to a minimum and providing better signal-to-noise ratios.

A complete television signal requires not only the luminance and chrominance components of the picture, but also line and frame synchronising information to enable the receiver to re-assemble the incoming signals into a coherent picture, and also one or more channels of sound to accompany the pictures. Teletext signals and various engineering test signals have also come to be regarded as an integral part of the television signal. To provide for all these requirements, a group of experts of the European Broadcasting Union, which represents all the major broadcasters in Europe, developed the basic IBA MAC system so as to include all the 'extras', and this has led to the introduction of a complete satellite television transmission system known as the C-MAC/Packet System.

The time-division multiplexing and compression techniques have been used to fit in a burst of digital data at the start of each MAC line, which is transmitted using direct digital modulation of the carrier during the line-blanking period. This data period carries line and frame synchronising information plus multiple sound channels and a considerable amount of data. Any, or indeed all, of the sound channels can be replaced by a data channel which can carry extra teletext signals, captions for the hard-of-hearing, notes on the vision programme, computer programs, or special messages for your particular receiver. Effectively, then, the transmitted signal consists of a time division multiplex of the three different parts. The analogue components (luminance and chrominance) are transmitted in time multiplex, and the digital parts of the signal are sent in a packet multiplex, as explained below.

Now let us look at the C-MAC/Packet signal (Fig. 29) in more detail, starting with the digital data period, since this is the first part of the signal to be transmitted. The first section of the data burst provides three synchronising signals, for line, frame, and colour sequence identification.

Line syncs are provided at the start of the data burst period in the form of one of two 6-bit words, each preceded by a run-in bit. The two 6-bit line-sync words alternate line by line except at the end of the frame where there is a break in the sequence so that odd and even fields can be identified.

Frame synchronisation information is also transmitted on line 625, immediately following the line sync words. There is a run-in period of 32 bits followed by a 64-bit frame synchronising word. This 96-bit sequence is transmitted in inverted form before odd-numbered frames, so if receivers detect this inversion they have a second method of achieving frame synchronisation.

Colour sequence identification is achieved by line counting from the frame sync signal. It has been arranged that odd lines of the frame carry the U (B-Y) colour difference signal, and that the V (R-Y) signal is carried on even lines. (See Ref. [6] for details of U, V). MAC uses a line-sequential colour system, but sophisticated vertical filtering produces much better results than would be obtained by simply omitting alternate lines or taking the average value of adjacent lines. The rest of the period consists of 198 bits of data in

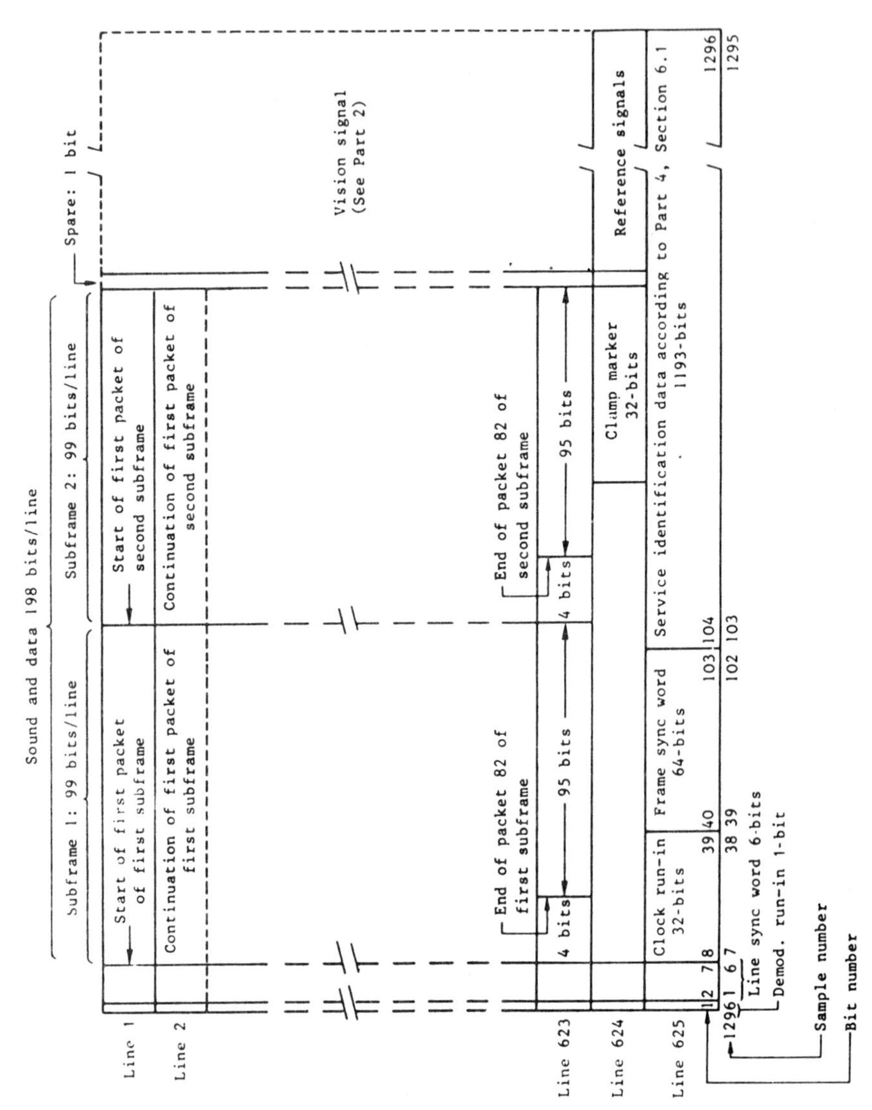

Sound and data 198 bits/line
Subframe 1: 99 bits/line
Subframe 2: 99 bits/line
Spare: 1 bit
Start of first packet of first subframe
Start of first packet of second subframe
Continuation of first packet of first subframe
Continuation of first packet of second subframe
Vision signal (See Part 2)
End of packet 82 of first subframe
End of packet 82 of second subframe
4 bits
95 bits
Clamp marker 32-bits
Reference signals
Clock run-in 32-bits
Frame sync word 64-bits
Service identification data according to Part 4, Section 6.1
1193-bits
Line 1
Line 2
Line 623
Line 624
Line 625
1296
1295
103 104
102 103
39 40
38 39
1 2 7 8
1296 1 6 7
Line sync word 6-bits
Demod. run-in 1-bit
Sample number
Bit number

Fig. 29

discrete packets, and it is this information that is used to carry the many different options of sound, teletext, or computer information.

Line 625, which, as we mentioned earlier, carries the frame synchronisation information, also carries in it a special data burst with information which tells the receiver how the incoming data bursts on each line are to be interpreted at any particular time. For example, whether 8 mono sound channels or 3 stereo sound channels plus data are being radiated.

To convert all the various digital sound and data signals into a single, continuous bit-stream, a form of multiplexing known as 'packet multiplexing' is adopted. Each packet (Fig. 30) has a constant length and consists of a string of 751 bits. The packet is divided into two parts, the header, and the data area. Every broadcast service has its own unique address code, carried in the header, which is used by the receiver to recognise and select the service required. Up to 1024 different simultaneous services could be offered.

The actual data area of the packet can carry 728 useful data bits of sound and control information.

It is instructive to think of the whole data burst section not as a certain number of sound channels or data channels, but as a data resource containing some 3 M bit/s of data, which can be used to provide whatever services are required. The flexibility that such an approach gives has tremendous advantages. The broadcaster had total control of the mixture of audio and data services. Instead of 8 high-quality sound channels, the broadcaster can choose to provide 16 lower bandwidth channels, or any other combination of sound and data within the limits of the capacity of the system. Outside normal broadcasting hours, when no sound is radiated, the whole time slot can be used to carry data, perhaps to provide thousands of pages of teletext-type information.

Moving along the C-MAC/(Packet) signal, let us look at the chrominance signals in a little more detail (Fig. 31).

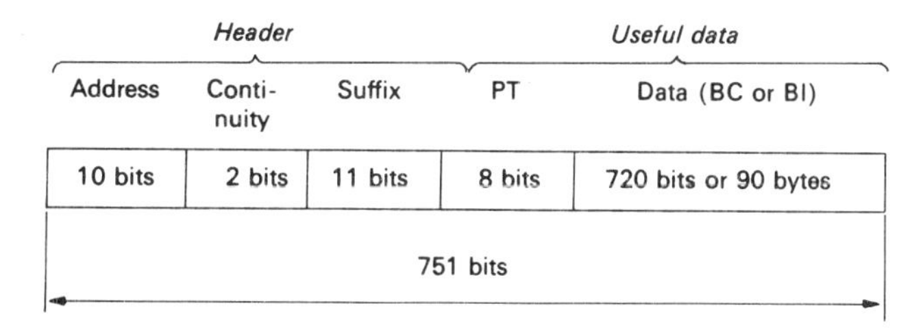

Fig. 30 – MAC packet structure.

The packet-type byte PT is inserted in packets carrying sound services to distinguish between coding blocks BC and interpretation blocks BI. For other types of services, the data area extends over 91 bytes. (EBU Review Technical)

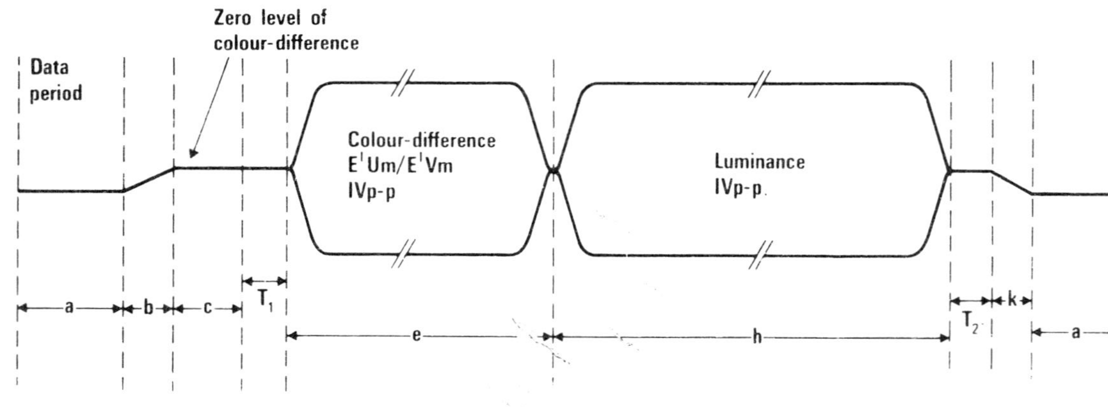

Fig. 31 – Detailed video waveform for C-MAC (Packet) picture transmissions (not to scale). Clock frequency: 20.25 MHz.

a = 206 bits for synchronisation, sound/data†
b = 4 clock periods for transition from end of data; includes leading edge of pedestal signal added to video to provide energy dispersal
c = 15 clock periods – clamp period (0.5 V)
T_1 = 10 clock periods which include a weighted transition to colour-difference signal of 5 clock periods
e = 349 clock periods for colour-difference component
T_2 = 5 clock periods for weighted transition between colour-difference and luminance signal
h = 697 clock periods for luminance component
T_3 = 6 clock periods for weighted transition from luminance signal
k = 4 clock periods for transition into data: includes trailing edge of pedestal signal to provide energy dispersal

† The complete data burst includes
1 run-in bit
6 bits of line sync word
198 bits of data (in two subframes of 99 bits each)
1 spare bit

Whenever you see MAC 'waveform' diagrams you will see that they are labelled, not by accurately defined timing information given in microseconds, but with 'numbers of samples' along the X-axis. The whole development of MAC as a system with a good deal of potential for the future, has been arranged so that the system ties in with the internationally agreed sampling frequencies for digital television, published as CCIR recommendation 601 (1982). The standard sampling frequencies are 13.5 MHz for luminance and 6.75 MHz for chrominance. These frequencies, when multiplied by the compression ratios used in MAC (3:1 for chrominance and 3:2 for luminance), give a sampling rate of 20.25 MHz.

It is often found helpful to consider the MAC waveform as an analogue waveform sampled at a rate of 20.25 MHz, effectively dividing the waveform into time slots 1/20.25 MHz or about 49.4 nanoseconds long. Since the length of one line of the C-MAC/Packet waveform is the standard 64 μs, we end up with $64/\frac{1}{20.25} = 1296$ samples per line, as Fig. 31 indicates.

Using this method of measurement, then our colour-difference signal occupies 349 sample periods, which is just about 17 μs. The maximum amplitude of the signal is 1 V peak-to-peak, and you will see from Fig. 31 that the signal is bipolar, so that zero chrominance occurs at mid-grey in luminance terms. The two chrominance signals U and V are transmitted on alternate lines.

Passing along to the luminance signal, you will see from the diagram that its amplitude is 1 V peak-to-peak with black at -0.5 V and white at 0.5 V with respect to the chrominance zero level. The luminance information occupies 697 samples, a time period of around 34 microseconds.

Since the MAC system was specifically designed to be used over FM satellite channels its various parameters have been chosen so as to make the best possible use of the available spectrum. The chrominance and luminance signal-to-noise ratios have been well matched to the characteristics of an FM channel, so that a significant improvement in noise performance is noticeable on picture. This, together with a complete lack of cross-effects and the flexible packet-sound/data system, was sufficient to make the European Broadcasting Union recommend that any member in Europe starting a 12 GHz satellite broadcasting service should use MAC, or more formally, the C-MAC/Packet system. The characteristics of the system are summarised in Table 4.1.

Table 4.1 Characteristics of the C-MAC/packet system

Baseband bandwidth	8.4 MHz
Luminance bandwidth	5.6 MHz
Compression ratios	
Luminance	3:2
Chrominance	3:1
Chrominance bandwidth (transmission)	~2.0 MHz

The chrominance performance of MAC has been arranged to be about 5 dB better than for PAL, with the luminance noise performance slightly worse, so as to give a good balance of noise performance across the whole video band.

4.3 APPLICATIONS OF MAC TO CABLE SYSTEMS

Designed from its inception as a satellite broadcasting system, the C-MAC/Packet system makes full use of the extended bandwidth permitted by the satellite channel. It was not, therefore, surprising that many cable service operators expressed views about the possibilities or impossibilities of passing C-MAC/Packet signals received from a satellite over some restricted bandwidth cable circuits. The EBU experts have therefore devised ways of ensuring that C-MAC/Packet signals can be recoded into the form of D-MAC/Packet signals (the D signifying that duo-binary data coding is used) (Ref. [8]), which will pass through cable networks with a channel bandwidth of 10.5 MHz. Also, since there are in Europe some existing cable services with channel bandwidths of as little as 7 MHz, a half bit-rate D2 MAC/Packet system has been specified to allow some received satellite signals to be used, after conversion, over even these narrow band circuits.

Inevitably, since the D2 MAC/Packet system can only cope with data at half the standard D MAC rate, fewer sound or data services can be provided; in addition, there is a small reduction in picture quality when signals using this system are passed through the narrow band cable distribution systems.

Great care has been taken to see that C, D, and D2 signals can be easily transcoded between the different systems, and the specialist integrated circuits that are being designed to decode MAC-type signals will be able to cope with any of the three systems without modification. At the time of writing (early 1985) final decisions on the small print of the specifications of the three systems have yet to be made, so the reader should not regard the details of the C-MAC/Packet system given earlier as unchangeable; small amendments may yet need to be made before the system comes into general use.

Other variants of MAC have been proposed and developed for use in particular circumstances. Canada and Australia have announced plans to use B-MAC, a system which uses video signals sensibly identical to those of C-MAC, but which carries the audio and data in a baseband multiplex using a multi-level code which allows a reduced bandwidth to be used. Although the noise performance of such signals is inferior to that of the C-MAC (Packet) system, and the sound/data capacity is reduced, B-MAC does provide a satellite signal that can be used without modification on cable systems.

REFERENCES AND BIBLIOGRAPHY

[1] Television systems for DBS. *The Radio and Electronic Engineer*, **52,** No. 7, July 1982.

[2] Compatible higher definition TV. *IBA Technical Review* 21.

[3] *Final Acts of the World Broadcasting Satellite Administrative Radio Conference*, Geneva 1977. ITU.

[4] CCIR Recommendation 601. *Encoding parameters of digital television for studios.*

[5] EBU Document SPB 284, June 1983. Rev. December 1984. *Television standard for the broadcast satellite service.* Specification of C-MAC/Packet System.

[6] *Specification of television standards for 625-line system I transmission in UK.* DTI 1984.

[7] H. Mertens & D. Wood. The C-MAC/Packet system for direct satellite television. *EBU Technical Review* August 1983.

[8] *Draft specification of the D and D/2 MAC/Packet systems.* EBU Document SPB 352.

5

Satellite link parameters

5.1 SATELLITE SLANT RANGE/BORESIGHT CALCULATION

In a previous section, equations were derived which permitted the calculation of velocity, period, or distance of a satellite in geostationary orbit. Of course, these equations are also applicable to satellites in a non-geostationary orbit i.e. for those in a polar orbit. Let us now calculate the slant range s for a geostationary satellite. In the first instance, consider the satellite ground station to be located at the *same* longitude as the geostationary satellite. From Fig. 32 it can be seen that the distance between the satellite S and the ground

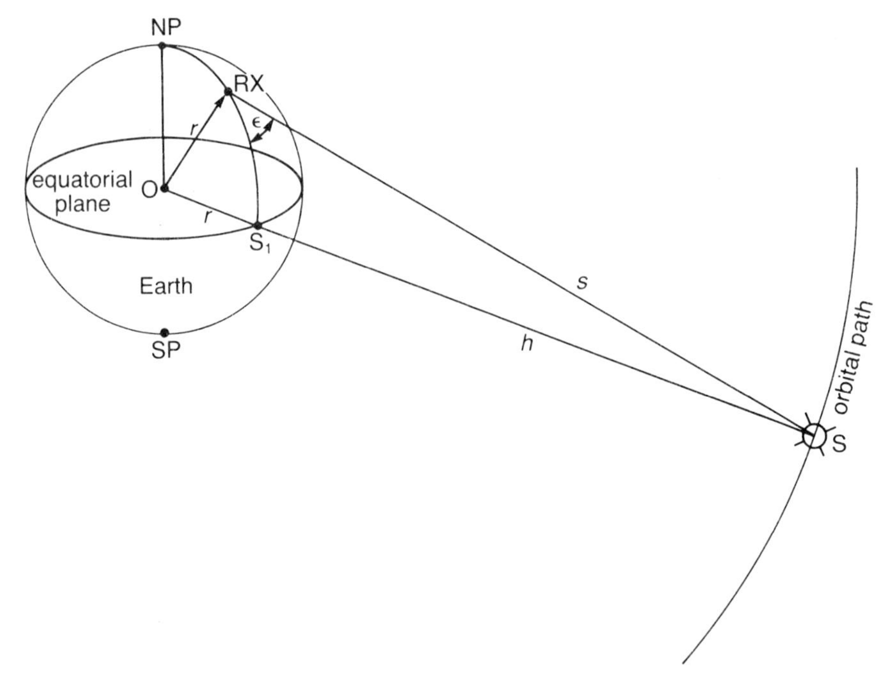

Fig. 32 – Geometry of geostationary satellite (S) in relation to a receiver RX on the same longitude.

station RX at S_1 is h, and is at its minimum. As the satellite receiving station RX moves further north along the line RX to NP, the distance s, also referred to as slant range, increases. The elevation angle ε of the receiver antenna decreases, eventually being zero. This is the case when the slant range s is tangential to the receiver location RX. Thus, there is a limit as to how far north or south one can receive signal transmissions from a geostationary satellite. For practical purposes the elevation must be larger than 5° to ensure adequate reception. The slant range will have to be taken into account when calculating the signal strength of a satellite link.

To assist understanding and to calculate the slant range, Fig. 32 has been re-drawn in Fig. 33. Using the law of sines one obtains

$$\frac{d}{\sin(\varepsilon + 90^\circ)} = \frac{s}{\sin \omega} = \frac{r}{\sin \sigma} \tag{5.1}$$

and

$$\sin \sigma = \frac{r \sin(\varepsilon + 90^\circ)}{d}. \tag{5.2}$$

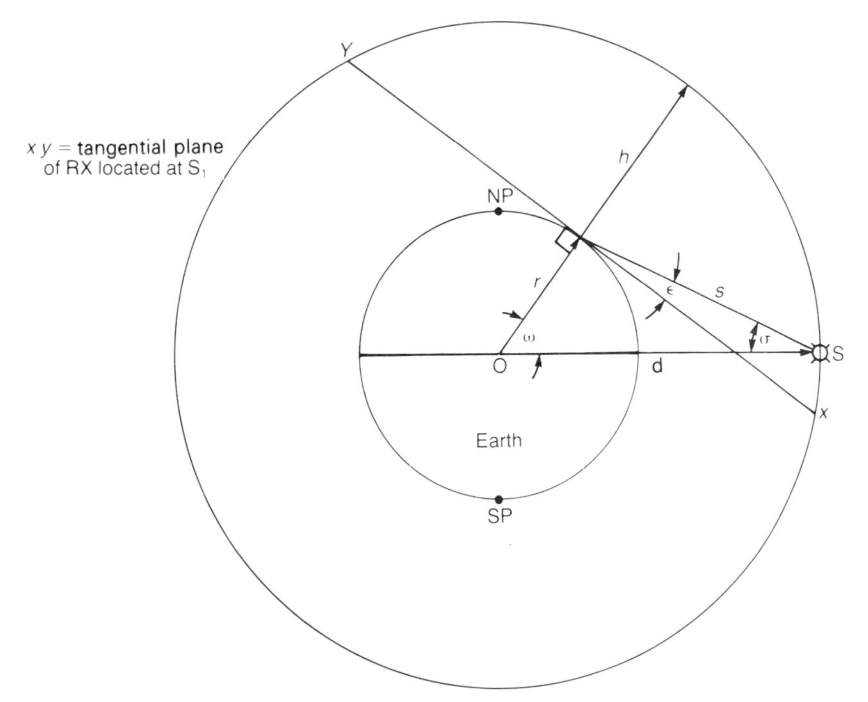

Fig. 33 – Satellite geometry for calculating slant range s.

The angle ω is also referred to as 'central angle'. Since $\omega+\sigma+(\varepsilon+90^\circ)=180^\circ$ the central angle can be calculated from

$$\omega = 90^\circ - (\varepsilon + \sigma) \tag{5.3}$$

where σ is given by equation (5.2).

From the law of cosines and with Fig. 33 one can calculate the slant range as

$$s = \sqrt{(r^2 + d^2 - 2\,rd\cos\omega)} \tag{5.4}$$

where

$$\omega = 90^\circ - \left[\varepsilon + \sin^{-1}\left(\frac{r\sin(\varepsilon + 90^\circ)}{d}\right)\right]. \tag{5.5}$$

The above equation can be further simplified by substituting the constant values for r and d. The slant range is thus simply a function of the antenna elevation angle.

Now let us consider the case where a geostationary satellite is received by a station RX which is *not* positioned on the same longitude as the satellite, as shown in Fig. 34a. It should be obvious that the slant range is at its minimum when the receiver is located at the sub-satellite point S_1, and that it increases when the receiver moves either eastwards or westwards. In addition, the slant range increases the further the receiver is moved to the north pole NP or south pole SP. From this, it can be seen that each satellite receiver scattered around the hemisphere requires a different aerial adjustment in terms of elevation and

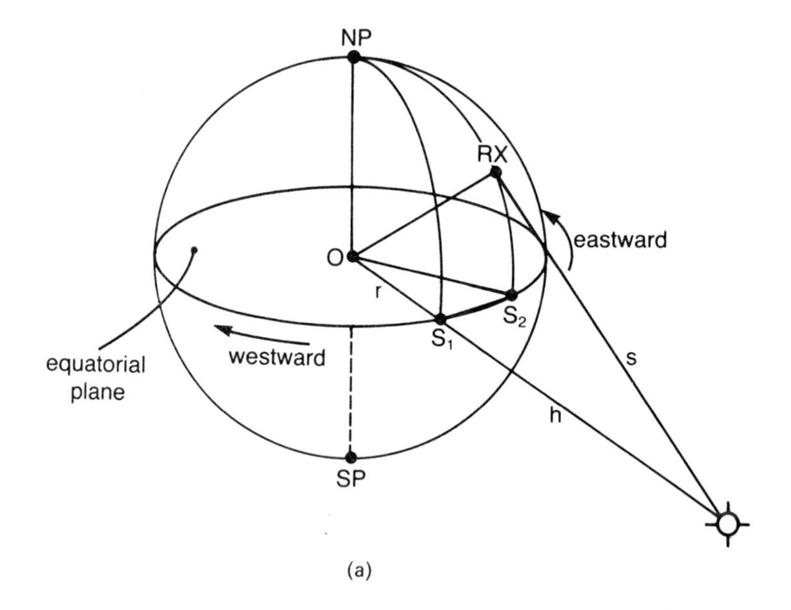

(a)

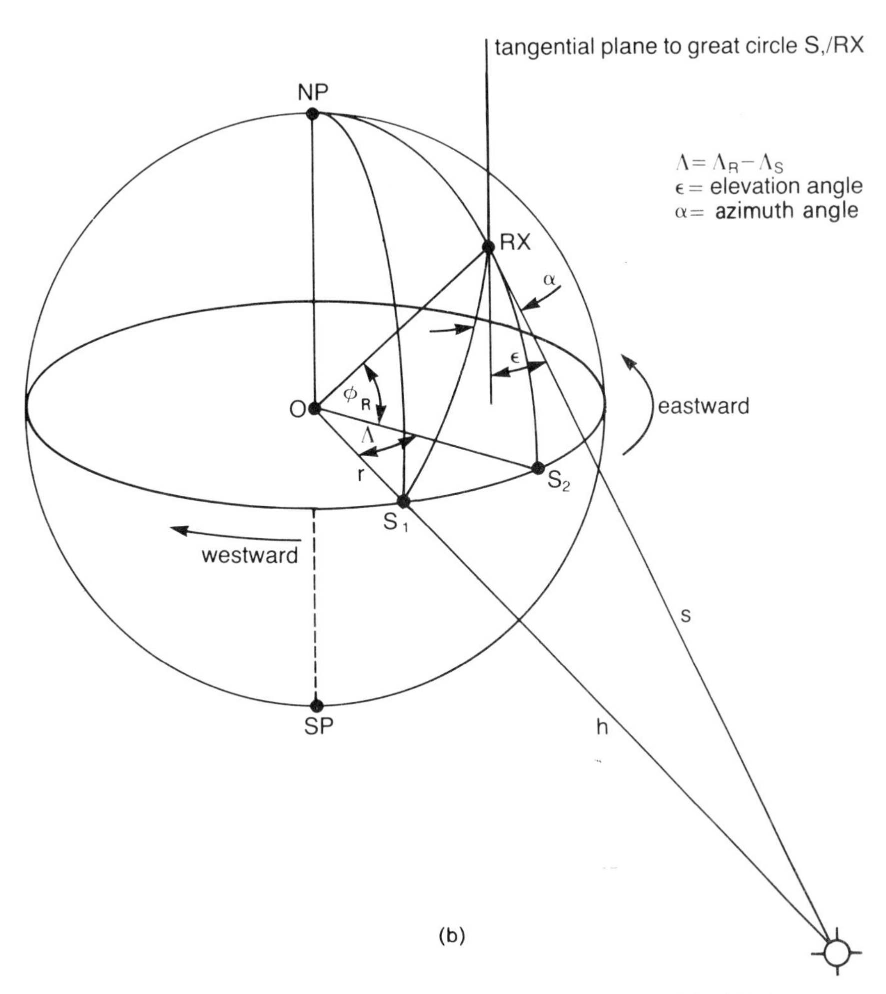

Fig. 34 – Geometry of geostationary satellite in relation to a receiver RX which is *not* on the same longitude as the satellite.

azimuth. If a receiver is located at longitude Λ_R and latitude ϕ_R, then the aerial elevation angle ε and azimuth angle α as well as the slant range s can be calculated from Fig. 34b, namely

$$\alpha = \text{arc tan}\,(\tan \Lambda/\sin \phi_R) + 180°, \tag{5.6}$$

where

$$\Lambda = \Lambda_R - \Lambda_S \tag{5.7}$$

and

$$\varepsilon = \text{arc tan}\,[(\cos \beta - r/d)/\sin \beta] \tag{5.8}$$

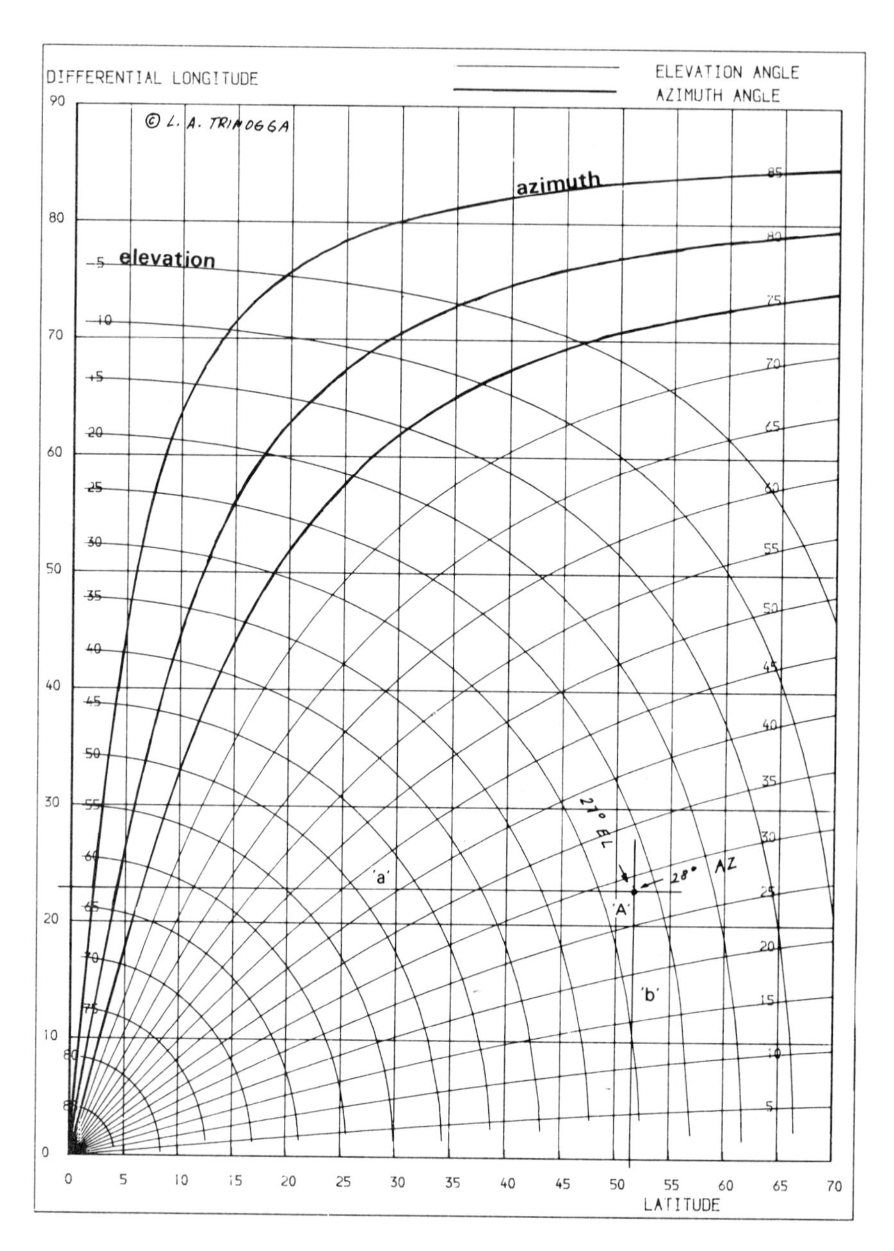

Fig. 35 – Nomogram for EL–AZ calculations.

where

$$\beta = \text{arc cos}\,(\cos\phi_R \cos\Lambda). \tag{5.9}$$

Substituting values for the equator radius and satellite height, the slant range can be found to be

$$s = 35779\sqrt{(1 + 0.42(1 - \cos\beta))}. \tag{5.10}$$

For Newcastle-upon-Tyne situated at 54.59°N, 1.35°W the receiver antenna elevation must be adjusted to 24.67° and the azimuth to 205.96° in order to receive signals from a geostationary satellite stationed at longitude 23°W. A programme, written in BASIC, can be found in the Appendix which permits the calculation of elevation, azimuth, and slant range. Alternatively, the elevation and azimuth angles can be obtained from the graph shown in Fig. 35. The procedure for finding the angles is as follows:

1. Note longitude and latitude of place where receiver is located. This information can be obtained from any good atlas; e.g. London 51.30°N/0.10°W.
2. Note longitude of satellite to be accessed, e.g. 23°W.
3. Calculate the difference in longitude between satellite and receiver, 23° − 0.1° = 22.9°, and enter in Fig. 35 as line 'a'.
4. Draw line 'b' at latitude 51.30° which is the latitude of the earth station. The two lines intersect at 'A'.
5. Read off angles of azimuth and elevation lines passing through 'A'. This is approximately 28° for azimuth and 27° for elevation.

5.2 SATELLITE COVERAGE

The area on earth which is covered by a geostationary satellite is egg-shaped with the sharp side of the egg pointing towards the pole. The size of the coverage area depends on the satellite aerial semi-beamwidth δ. With the help of Fig. 36 the coverage radii E and hence the approximate coverage area can be calculated from:

$$E_1 = a_1 + \frac{{a_1}^2}{2[r\tan(\varepsilon - \delta) - a_1]} \tag{5.11}$$

$$E_2 = a_2 - \frac{{a_2}^2}{2[r\tan(\varepsilon + \delta) + a_2]} \tag{5.12}$$

$$E_3 = s\tan\delta, \tag{5.13}$$

where

$$a_1 = \frac{s\tan\delta}{\sin(\varepsilon - \delta)} \quad \text{and} \quad a_2 = \frac{s\tan\delta}{\sin(\varepsilon + \delta)}. \tag{5.14}$$

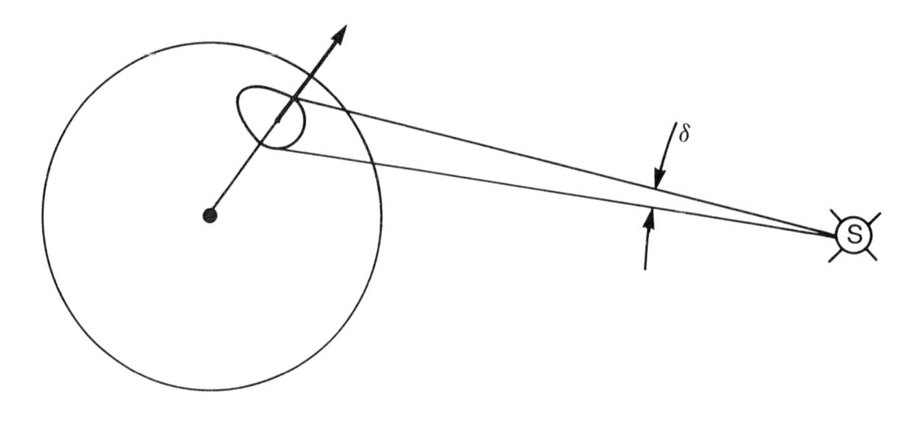

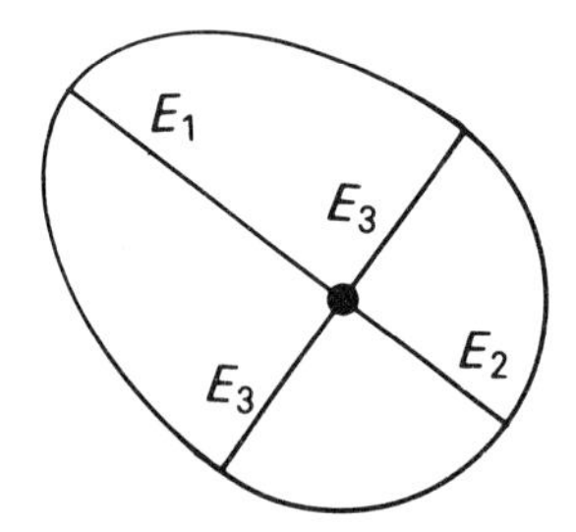

Fig. 36 – Satellite coverage: beamwidth and coverage shape.

Problem 1

Calculate the parameters relating to a DBS in ideal geostationary orbit. Hence show that a drift of one degree per day relates to an orbital difference in distance of 80 km.

Answer

$$T^2 = 4\pi^2 \frac{(r+h)^3}{G m_e}$$

$$(r+h)^3 = d^3 = \frac{T^2 G m_e}{4\pi^2} = (86164^2 \times 6.67 \times 10^{-11} \times 5.977 \times 10^{24})/4\pi^2$$

$$= \frac{295.8 \times 10^{22}\ \text{m}^3}{4\pi^2} = 7.4927 \times 10^{22}\ \text{m}^3$$

$$\underline{d = 4.215 \times 10^7\ \text{m} = 42\,157\ \text{km}}$$

$$h = d - r = 42\,157 - 6378 = \underline{35\,779\ \text{km}}$$

$$v = \frac{2\pi d}{T} = \frac{2\pi \times 42\,157}{86\,164} = \underline{3\,074 \text{ km/s}}$$

86 164 s = 360° therefore 1° = 239.3444 s.

let now $T = T - 239.3\,444$ s

$$d^3 = \frac{T^2 G m_e}{4\pi^2} = \frac{(86164 - 239.3444)^2 G m_e}{4\pi^2}$$

$$d = 42.077 \text{ km}$$

$$d_{\text{radius error}} = d_{\text{perfect}} - d_{1^\circ} = 42.157 - 42.077$$

$$\underline{d_{\text{radius error}} = 80 \text{ km}}$$

Problem 2

Calculate the slant range s for a DBS, assuming a receiver aerial elevation of 90°. Repeat the calculation for an aerial elevation of 26°.

Answer

$$r = 6378 \text{ km} \quad d = r + h = 42\,157 \text{ km}$$

$$\text{for} \quad \omega = 0^\circ \quad d = r + s = r + h \quad \text{or} \quad s = h$$

Calculate the slant range for $\omega = 0°$. This means $\varepsilon = 90°$ from Fig. 33.

From (5.5)

$$\varepsilon = 90^\circ - (90^\circ + 0^\circ) = 0^\circ$$

$$\cos\omega = \cos 0^\circ = 1$$

$$s = \sqrt{(r^2 + d^2 - 2r\,d)}$$

$$s = \sqrt{((0.4067 + 17.77 - 5.377) \times 10^8)}$$

$$\underline{s = 35\,776 \text{ km}}$$

$$\text{EL} = \varepsilon = 26^\circ \quad h = 35\,779 \text{ km} \quad r = 6378 \text{ km}$$

$$\varepsilon + 90^\circ = 116^\circ \quad \sin 116^\circ = 0.898 \quad \frac{r \sin 116^\circ}{d} = 0.1358$$

$$\sin^{-1} 0.1358 = 7.8^\circ$$

$$\varepsilon + 7.8^\circ = 33.8^\circ \quad \omega = 90^\circ - 33.8^\circ = 56.2^\circ \quad \cos\omega = 0.556$$

$$2rd \cos\omega = 5.377 \times 10^8 \times 0.556 = 2.99 \times 10^8 \text{ km}$$

$$s^2 = (0.4067 + 17.77 - 2.99) \times 10^8$$

$$\underline{s = 38\,970 \text{ km}}$$

Problem 3

A geostationary satellite is positioned at longitude 23° W. Calculate azimuth, elevation, slant range, and coverage radii for the following European cities for a semi-bandwidth of 1°:

Newcastle-upon-Tyne	54.59° N/1.35° W
Cologne	50.56° N/6.57° E
Belfast	54.35° N/5.55° W
London	51.30° N/0.10° W
Cape Town	33.56° S/18.28° E

Answer

	ε [°]	α [°]	s [km]	E_1 [km]	E_2 [km]	E_3 [km]
Newcastle-upon-Tyne	24.6	205.9	39 100	3 012	1 308	682
Cologne	25.6	216.3	39 004	2 644	1 272	680
Belfast	25.9	201.1	38 979	2 569	1 263	680
London	27.4	208.4	38 842	2 223	1 215	677
Cape Town	31.3	237.8	38 492	1 697	1 107	671

Problem 4

Calculate the boresight for a 12 GHz receiver located in Newcastle-upon-Tyne at 54.59°N, 1.35°W. Assume a semi-beamwidth of 1° and a satellite position of 23°W. Calculate all relevant parameters using the computer program 'BORESIGHT' listed in Appendix.

Repeat the calculation for Leeds.

Answer

```
BORESIGHT calculations

l=ls-lr   (■)        21.65
azimuth  (■)         205.96721
elevation  (■)       24.675667
pathl'gth d(km)      39092.635
frequency(GHz)       12

free space loss      -205.8673 dB

E1                   3010.8672

E2                   1307.8984

E3                   682.36448

delta                1
```

```
BORESIGHT calculations

l=ls-lr    (°)        21.65
azimuth  (°)          206.27978
elevation  (°)        25.722662
pathl'gth d(km)       38994.085
frequency(GHz)        12

free space loss  -205.84537 dB

E1                    2636.1877

E2                    1271.3847

E3                    680.64431

delta                 1
```

Problem 5

Calculate the free space loss for a receiver working at 12 GHz which is located at the sub-satellite point of a geostationary satellite. Cross-check using the computer programme 'BORESIGHT'.

Answer

$d = h$, i.e. shortest distance possible between RX and satellite, therefore

$$h = 35\,779 \text{ km}$$

$$\alpha = 10 \log\left(\frac{\lambda}{4\pi d}\right)^2 \quad \lambda = \frac{c}{f}$$

$$= 10 \log\left(\frac{c}{f 4\pi d}\right)^2$$

$$= 10 \log\left(\frac{3 \times 10^{10}}{12 \times 10^9 \times 4 \times \pi \times 3.5779 \times 10^9}\right)^2$$

$$= 10 \log (5.56 \times 10^{-11})^2$$

$$= -205.09 \text{ dB}$$

```
BORESIGHT calculations

l=ls-lr    (°)        0
azimuth  (°)          180
elevation  (°)        89.998543
pathl'gth d(km)       35779
frequency(GHz)        12

free space loss  -205.09796 dB

E1                    625.15507

E2                    625.15296

E3                    624.52477

delta                 1
```

Problem 6

A receiver designed for the reception of signals from OTS at 11.64 GHz is located at 54.59°N/1.35°W. Calculate the free space path loss for this satellite link. OTS is positioned at 10°E.

Answer

$\alpha = -205.55$ dB
$d = 38\,888.94$ km (from spectrum computer program)

```
BORESIGHT calculations

l=ls-lr      (°)        11.35
azimuth      (°)        193.83564
elevation    (°)        26.85293
path length d(km)       38888.944
frequency(GHz)          11.64

free space loss         -205.55735 dB

E1                      2344.5143

E2                      1234.2749

E3                      678.80904

delta                   1.
```

5.3 RADIO PROPAGATION IN SPACE–EARTH LINKS

In satellite systems, electromagnetic waves form the link between the satellite and earth station [1, 2]. During their travel, these waves are subject to physical laws over which man has no control. Thus, in order to establish a high-quality satellite link, it is important to have some understanding of the different factors which affect wave propagation and which permit the calculation of a satellite link.

For a given specification of a complete satellite system it would be ideal if one could predict the exact signal strength at any point along the path of propagation. Unfortunately, there are a number of factors which interfere and make such a prediction difficult. To begin with, the medium through which electromagnetic waves pass has properties which vary with temperature, water vapour, and pressure, which in turn alter the direction, polarisation, and velocity of radio waves. Besides the transmission medium, it is also the characteristic of the ground and space station which has a bearing on the quality of the satellite link. By characteristic, one means the frequency used, the type of aerial system, and the type of modulation employed. It is the task of the engineer to find and use the propagation conditions and electronic circuits which are most suitable to a high-quality satellite link.

There are various factors which have to be considered when choosing a frequency for satellite transmissions. The most obvious factor is to choose a high enough frequency to obtain large bandwidth consistent with transmission requirements. The second factor is choosing a frequency related to the frequency-dependent atmospheric attenuation. This is discussed in one of the following sections. Thirdly, the natural noise entering the satellite link via the sky, the so-called 'sky noise', has to be taken into account. These points are now briefly discussed.

To obtain an expression for the satellite down-link (propagation through free space) consider the arrangement shown in Fig. 37. A high-power amplifier HPA feeds the aerial, which has a gain G_{aTX}, with a power P_a. The maximum radiated power at the output of the aerial is thus the product of the aerial gain and the power applied to it, or

$$P = P_a + G_{aTX} \tag{5.15}$$

if all quantities are expressed in dB. The signal level between two points varies inversely with the square of the distance between these two points. For an isotropic aerial the electric field strength E at a distance d from the source of power will be

$$E = \frac{P}{4\pi d^2} \tag{5.16}$$

In section 5.6 the gain of a parabolic dish aerial is given by

$$G = \frac{4\pi\eta A}{\lambda^2} \tag{5.17}$$

Hence

$$A = \frac{G\lambda^2}{4\pi\eta} \tag{5.18}$$

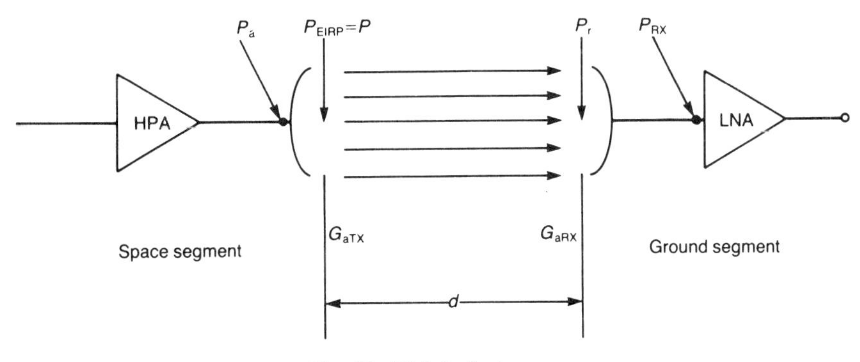

Fig. 37 – Link budget powers.

If an aerial is placed in the satellite signal path, then the power level P_r received by that aerial is

$$P_r = E \times A = \frac{P}{4\pi d^2} \times \frac{G\lambda^2}{4\pi\eta}. \tag{5.19}$$

For an isotropic aerial $G = 1$, and

$$\frac{P_r}{P} = \left(\frac{\lambda}{4\pi d}\right)^2 = \alpha \tag{5.20}$$

where α is known as the 'free space path loss' or simply as free space loss. This is usually expressed as a power ratio in dB:

$$\alpha = 10 \log\left(\frac{\lambda}{4\pi d}\right)^2. \tag{5.21}$$

The power P_{RX} is available at the receiver input is

$$P_{RX} = P_r \times G_{aRX} \tag{5.22}$$

or

$$P_{RX} = P_a G_{aTX} \alpha G_{aRX} \tag{5.23}$$

or

$$P_{RX} = P_a + G_{aTX} + \alpha + G_{aRX} \tag{5.24}$$

where all parameters in equation (5.24) are expressed in dB. For example, the free space path loss at the sub-satellite point of a geostationary satellite working at 12.1 GHz is 205.09 dB.

In Table 5.1 the free space path loss is given for a receiver positioned at latitude ϕ and at a relative longitude Λ with respect to the satellite. A frequency of 12.1 GHz was chosen from the frequency band 11.7–12.5 GHz for the purpose of calculation.

Table 5.1 Free space path loss for 12.1 GHz

Latitude ϕ (°) \ Relative longitude Λ (°)	0	10	20	30	40	50	60	70
0	205.09	205.12	205.20	205.33	205.50	205.70	205.92	206.15
10	205.12	205.14	205.22	205.35	205.52	205.71	205.94	206.17
20	205.20	205.22	205.30	205.42	205.57	205.76	205.96	206.18
30	205.33	205.35	205.42	205.52	205.66	205.83	206.02	206.21
40	205.50	205.52	205.57	205.66	205.78	205.93	205.09	206.26
50	205.70	205.71	205.76	205.83	205.93	206.05	206.18	206.32
60	205.92	205.93	205.96	206.02	205.09	206.18	206.28	206.39
70	206.15	206.16	206.18	206.21	206.26	206.32	206.39	206.46

Now let us look at aspects of wave propagation through the atmosphere. In the simplest case the atmosphere can be divided into the ionised and non-ionised regions. The altitudes to which these regions extend are approximately 60 km and 1000 km respectively, as can be seen from Fig. 38. It is the lower atmospheric regions which have the greatest influence on the propagation of radio waves from a space–earth link, because most meteorological phenomena take place here, resulting in continuous changes of temperature, humidity, pressure, clouds, and rainfall.

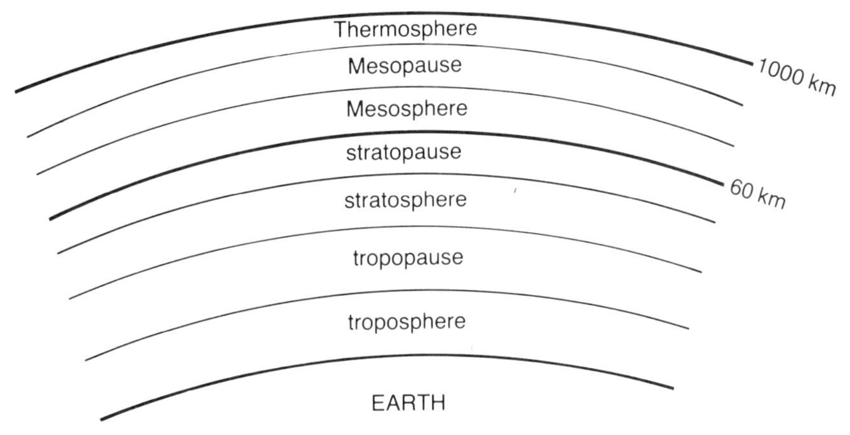

Fig. 38 – Possible division of atmosphere.

When electromagnetic waves pass through the atmosphere they are subject to attenuation. The longer the path through the atmosphere, the larger the attenuation. In other words, satellite aerials with low elevation angles face a larger loss than those with higher elevation angles. Attenuation is also frequency dependent. Satellite broadcasts at 12 GHz and 20 to 30 GHz are markedly affected by rainfall. This must be taken into account when planning a link in order to guarantee satisfactory operation under worst conditions. Furthermore, radio waves are depolarised to various degrees whilst travelling through the atmosphere. The total effect of the lower atmosphere on radio wave propagation is illustrated in Fig. 39.

Satellite links for direct broadcasting have to be designed conservatively in order to retain the satisfaction of the customer. A television viewer, for example, would not be very happy if the picture would suddenly degrade or even disappear because of a rain storm. For this reason a DBS link is calculated for satisfactory operation for 99 % or 99.9 % for the country or continent in question. Various organisations, including the IBA, have undertaken extensive propagation studies in the 11.7–12.5 GHz band. For a midband frequency of

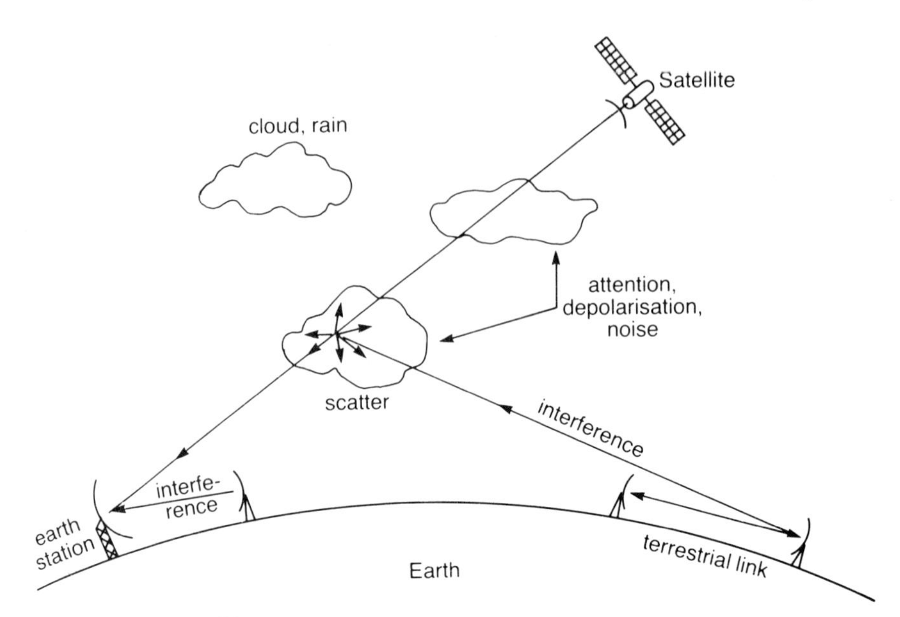

Fig. 39 – Effect of lower atmosphere on radio waves.

12.1 GHz the atmospheric attenuation for Western Europe using the least favourable month of the year is shown in Table 5.2. For example, for a satellite elevation of 25° (southern area of UK) the atmospheric attenuation is, in the worst case, 2.1 dB for 99 % of the time during which satellite broadcasts are received. If the time is increased by 0.9 % to 99.9 %, then this will result in a maximum atmospheric attenuation of 6.8 dB during some stage of the

Table 5.2 Atmospheric attenuation for a DBS midband frequency of 12.1 GHz. (percentage relates to total satellite broadcast time)

Satellite elevation angle (°)	Attenuation in dB not exceeding a time of	
	99 %	99.9 %
5	6.8	14
10	4.7	9.7
15	3.2	7.4
20	2.5	6.4
25	2.1	5.8
30	1.8	5.3
35	1.7	5.1
40	1.6	4.9
45	1.5	4.8

reception. Taking the free space path loss for Western Europe as 205 dB (see Table 5.1), then the total path loss would be 211.8 dB. For the remaining 0.1 % of the time, attenuation may increase another 8 dB to about 220 dB.

Table 5.2 relates to Western Europe and the least favourable month of the year for satellite reception. The figures quoted are based on experimental results obtained by ESA.

5.4 SKY NOISE

Communication systems such as those used for DBS, satellite communications, radio-astronomy, and radar, are placed or sandwiched between the earth and the sky. The earth and the sky are sources of noise, and this noise enters the communications receivers via its aerial and has thus to be taken into account in system design.

The noise, emanating externally to the earth, sometimes referred to as 'sky noise', can be attributed to various sources. Generally speaking the sky noise comes from the galaxy, the sun, the moon, radio stars, and other planets or discrete celestial bodies. Up to about several 100 MHz atmospheric noise is the main source of sky noise. Thereafter, and up to the region of 1 GHz, the main sources of noise are the galaxy and the sun. This contribution may be summarised as galactic noise.

The amount of noise received by a ground based receiver from the earth is given by Planck's law, namely

$$P = \frac{\varepsilon h f \mathrm{d}f}{e^{\frac{hf}{kT}} - 1} \tag{5.25}$$

where P = noise power radiated by a body
ε = emissivity
T = absolute temperature (K)
$\mathrm{d}f$ = bandwidth (Hz)
f = centre frequency (Hz)
h = Planck's constant 6.624×10^{-34} J s.

In other words, the noise power radiated by a body, like the earth, at a temperature T, emissivity ε, and bandwidth $\mathrm{d}f$, is P. If the earth is considered a black body of $\varepsilon = 1$, and if one considers a typical microwave link for which $hf \ll kT$, then equation (5.25) simplifies to

$$P = kT\,\mathrm{d}f. \tag{5.26}$$

The amount of noise power received from the earth by a ground based receiver can be reduced through the use of directional aerials, and depends on the latter's elevation. For low elevation the radio waves have to travel a long path through the atmosphere. For 90° elevation (pointing at the zenith) the path and hence the noise from the atmosphere are smallest. Most frequently the noise is expressed as 'effective noise temperature' T_e rather than in absolute power. Thus

$$T_e = \frac{P}{k\,df} \tag{5.27}$$

where P is the sky noise or noise power from any other source. The aerial noise temperature due to the atmosphere follows the general pattern shown in Fig. 40. This figure exhibits two peaks which are due to the water and oxygen absorption bands in the atmosphere. Below several hundred MHz the aerial noise temperature becomes negligible.

The noise coming from beyond the atmosphere, i.e. the galaxy, decreases very quickly between 100 MHz and about 1 GHz, as shown in Fig. 41. When the noise contribution from Fig. 40 and Fig. 41 are added one obtains the familiar graph for the sky noise temperature as shown in Fig. 42. The frequency bands of low noise temperature are referred to as 'atmospheric

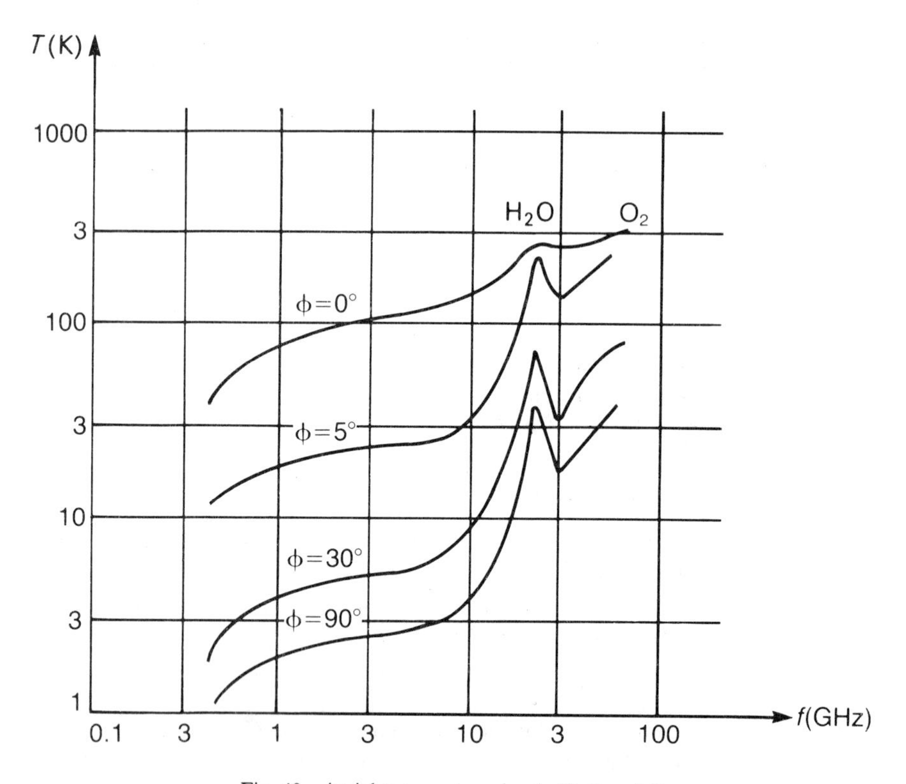

Fig. 40 – Aerial temperature due to H_2O and O_2.

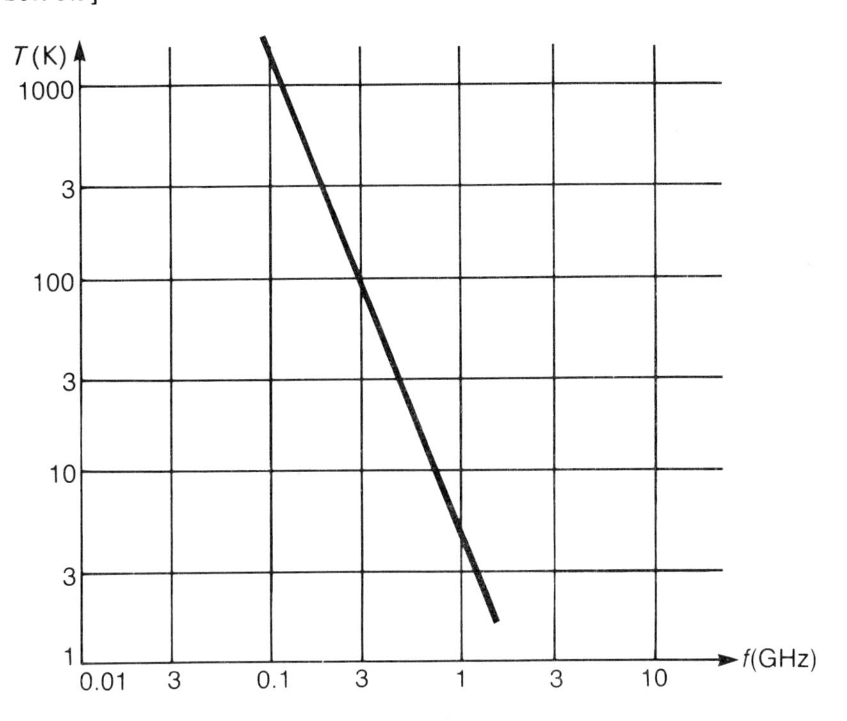

Fig. 41 – Mean sky noise temperature.

windows'. Figure 42 shows that such windows exist around 4, 30, and 80 GHz, and are thus suitable for satellite transmissions.

For the purpose of comparison the noise temperatures of various amplifiers, as quoted in the literature, are superimposed on Fig. 42.

5.5 NOISE

For economic reasons, signal levels in communication systems reach such low levels, that the effect of natural noise becomes apparent. Any additional noise due to unsuitable components or circuit design deteriorate the overall performance of the information system. The situation is particularly critical where the front end of a system is concerned.

In practice, two sources of noise [3] are encountered in satellite systems: 'natural noise' and 'man-made' noise. The latter can be attributed to such spurious voltages as hum and interference. Man-made noise can often be eliminated either by careful circuit design, screening, or rearrangement of components. Natural noise, however, cannot be eliminated, only minimised by proper circuit design and component selection.

As a signal passes through passive parts of the communication system

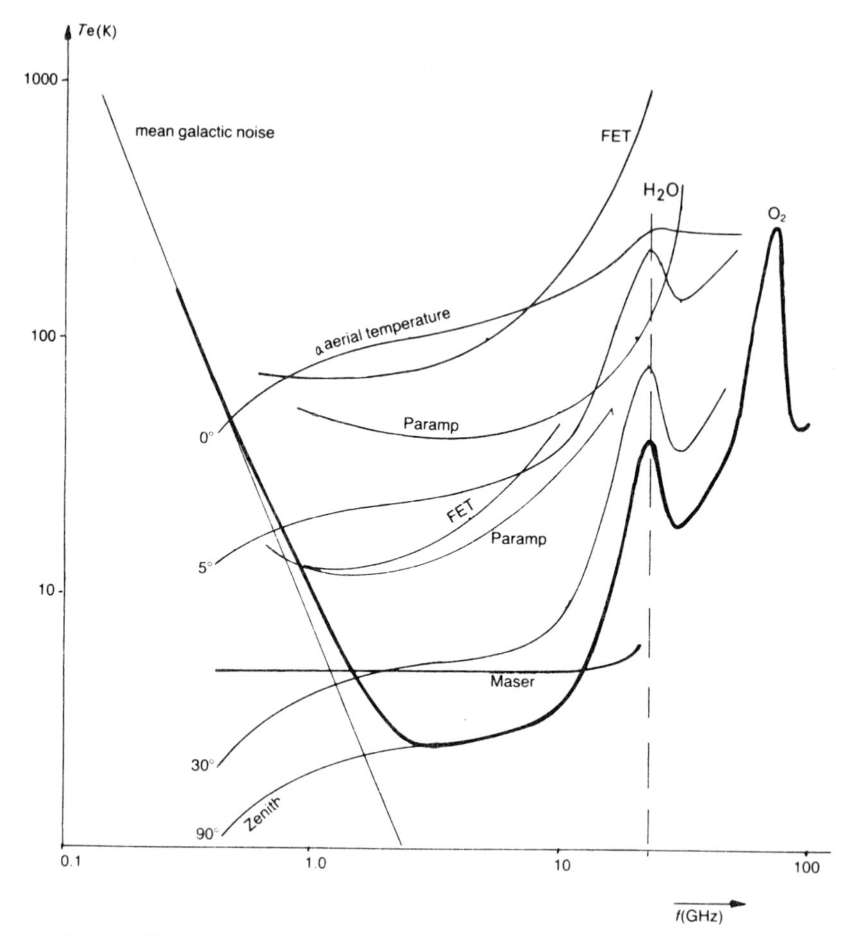

Fig. 42 – The total background noise for a DBS receiver (——) for different aerial elevation angles.

such as waveguides and cables it is attenuated. Although the signal is subsequently amplified to boost signal level, additional noise is then introduced by the amplifiers. To obtain a measure for the 'noisiness' of a system or at different points within the system, various concepts can be used. In the following the concepts of 'noise figure' and 'noise temperature' are explained, as relevant to satellite systems. More general information can be found in the appropriate literature.

5.5.1 Noise figure

A possible classification of noise is given in Fig. 43.

Thermal noise is quantified by the well-known equation

$$\overline{e^2_{\text{th}}} = 4kTR\,B \tag{5.28}$$

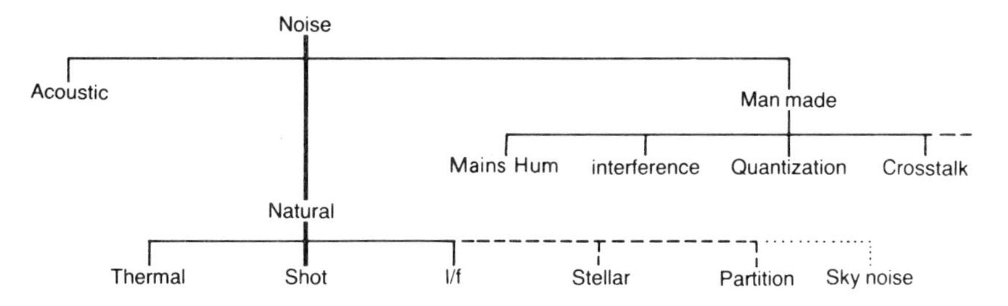

Fig. 43 – Possible classification of noise.

where k = Boltzman's constant 1.38×10^{-23} W s/K
$T = 290$ K
B = system bandwidth or spot frequency
R = source resistance.

Here we are considering only matched systems, i.e. where the source resistance is equal to the load resistance. Under this condition we obtain maximum power transfer. Applied to equation (5.28), the maximum thermal noise power available from a resistor is thus

$$\frac{\overline{e^2_{\mathrm{N}}}}{4R} = N = kTB. \tag{5.29}$$

The noise power N is thus directly proportional to temperature and bandwidth. The noise power per unit bandwidth is given by

$$N = kT = 4 \times 10^{-21} \text{ W/Hz} \tag{5.30}$$

if $T = 290$ K is taken as the standard reference temperature. For the same temperature the noise power may also be expressed as -204 dB W/Hz or -174 dB m/Hz. Together with the signal power S the signal-to-noise ratio (SNR) can be calculated at any point in a system. In analogy with equation (5.29) the signal power S is

$$S = \frac{e_{\mathrm{s}}^{2}}{4R}. \tag{5.31}$$

From this and with Fig. 44 we obtain for the signal-to-noise ratio at the input of a system, e.g. an amplifier,

$$\text{SNR} = \frac{S}{N} = \frac{e_{\mathrm{s}}^{2}/4R}{kTB}. \tag{5.32}$$

This is the best possible SNR one can achieve. All subsequent ratios will be worse, as is outlined in the following section.

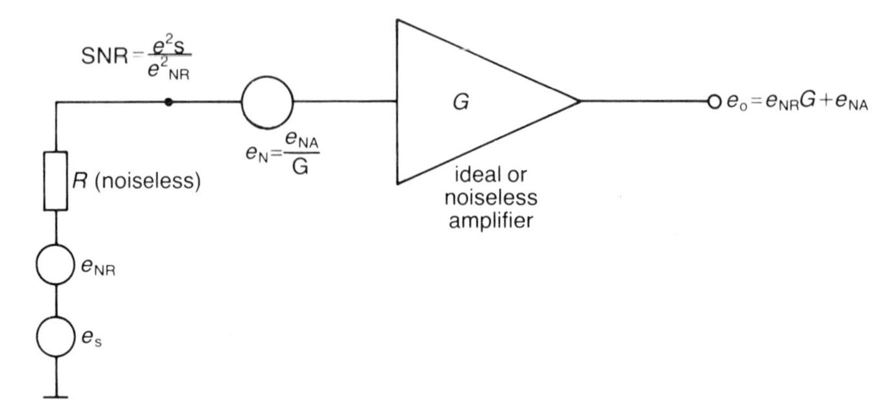

Fig. 44 – Equivalent circuit of a noiseless amplifier.

From Fig. 44 it is evident that the output noise for a noiseless amplifier is given by

$$e_o = e_{NR}G \tag{5.33}$$

where e_{NR} is the noise voltage from the source resistor **R**, and **G** the voltage gain of the amplifier. A practical amplifier, however, will generate noise which appears as a voltage e_{NA} at its output. For convenience this voltage may be referred to the amplifier input as e_{NA}/G. The total output noise voltage e_o of the amplifier is

$$e_o = e_{NR}G + e_{NA} = (e_{NR} + e_n)G. \tag{5.34}$$

To obtain a measure of the noise generated in a system or circuit, the concept of the noise figure is used. By definition this is the SNR at the input of a system to the SNR at its output, i.e.

$$F = \frac{\text{(SNR) input}}{\text{(SNR) output}} = \frac{\text{ideal SNR}}{\text{actual SNR}} \tag{5.35}$$

or

$$F = \frac{e_s^{\,2}/\overline{e_{NR}^{\,2}}}{e_s^{\,2}/(\overline{e_{NR}^{\,2}} + \overline{e_N^{\,2}})} = \frac{\overline{e_{NR}^{\,2}} + \overline{e_N^{\,2}}}{\overline{e_{NR}^{\,2}}}. \tag{5.36}$$

The noise figure is usually expressed in decibels

$$F\,[\text{dB}] = 10 \log F. \tag{5.37}$$

A noiseless circuit has thus a noise figure of one or zero decibel. For real circuits F is always larger than this value. To obtain the overall noise figure of a system the signal-to-noise ratios or noise figures of the sub-units need to be known, since each sub-unit generates its own noise.

For a cascade of several stages one obtains the well-known equation

$$F = F_1 + \frac{F_2 - 1}{G_1} + \frac{F_3 - 1}{G_1 G_2} + \ldots . \tag{5.38}$$

where F = overall or total noise figure
F_1 = noise figure of first stage
F_2 = noise figure of second stage, etc
G_1 = power gain of first stage
G_2 = power gain of second stage, etc.

The proof of equation (5.38) for a cascade of two amplifiers is left to the reader.

To illustrate the use of this equation consider two satellite receiver RF amplifiers as shown in Fig. 45. The first amplifier has a gain of 8 dB. As a numerical value this is

$$G_1 = \text{antilog}\,\frac{8}{10} = 6.30.$$

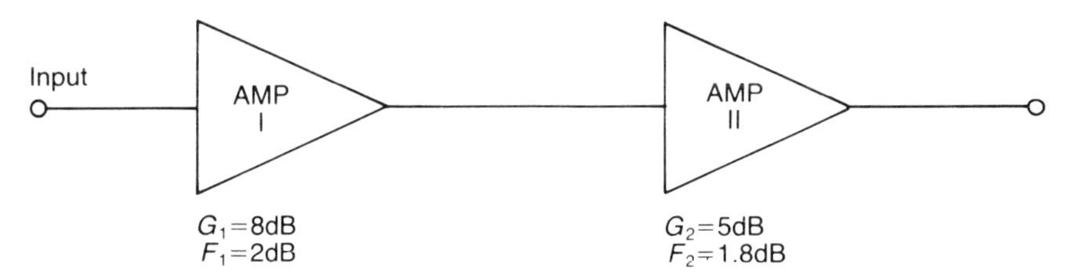

Fig. 45 – Cascade of two amplifiers.

Applying the same technique to the other parameters, the overall noise figure of the amplifier chain is found to be

$$F = 1.58 + \frac{1.51 - 1}{6.30} = 1.66 \text{ or } 2.20\,\text{dB}.$$

If the two amplifiers are interchanged, the overall noise figure becomes 2.28 dB. This shows that the order of the amplifiers does have a bearing on the overall noise figure. A step-by-step solution is given at the end of this chapter.

5.5.2 Noise temperature

In the previous noise analysis a constant temperature of 290 K was assumed for all circuits or sub-units within a system. This assumption is not true for satellite communication systems where the input unit, the aerial, points into the sky. The noise temperature of the sky is usually much lower than absolute temperature, and furthermore, is frequency dependent. The principle of noise temperature is useful for calculating the noise performance of a system at any source temperature, and is therefore used in radio astronomy and satellite work.

From equation 5.29, the noise temperature T_n is defined as

$$T_n = \frac{N}{kB}[K] \tag{5.39}$$

or per unit bandwidth as

$$T_n = \frac{N}{k}[K] = 7.25 \times 10^{22} \times N \tag{5.40}$$

with the dimensions of N as W/Hz.

For a noise power of 4×10^{-21} W/Hz we obtain a noise temperature of 290 K. A noise power of 3.2×10^{-22} W/Hz results in a noise temperature of 23 K.

The noise temperature thus relates the noise power available at the source of a system with the temperature in K of a resistor from which the same thermal noise power is available. Alternatively, the concept of equivalent noise temperature may be used. This is explained by means of two amplifiers in cascade, as shown in Fig. 46.

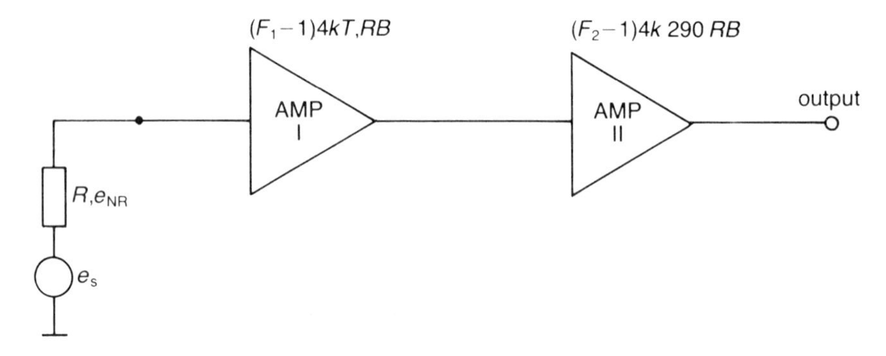

Fig. 46 – Two cascaded amplifiers.

From equation (5.36) one can express the noise figure as

$$F = 1 + \frac{\overline{e_N^{\,2}}}{\overline{e_{NR}^{\,2}}} = 1 + \frac{\overline{e_N^{\,2}}}{4kTRB}$$

or

$$\overline{e_N^{\,2}} = (F_1 - 1)4kT_1RB \tag{5.41}$$

where e_N is the equivalent input noise voltage. The equivalent input noise voltage of the second amplifier is

$$(F_2 - 1)4kT_2RB = (F_{290} - 1)4k\,290\,RB$$

if a reference temperature of 290 K is chosen. The noise voltage delivered by the first amplifier is $4kT_1RB$ where T_1 is the equivalent noise temperature of the first amplifier, which may not be at 290 K. The signal-to-noise ratio at the input of the second amplifier is thus

$$\frac{S}{N} = \frac{e_s^{\,2}}{(F_{290} - 1)4k290RB + 4kT_1RB} \tag{5.42}$$

or

$$\frac{S}{N} = \frac{e_s^{\,2}}{4k\,RB\,(T_e + T_1)}. \tag{5.43}$$

From this we obtain the relationship between noise figure and effective noise temperature as follows:

$$T_e = (F_{290} - 1)290 = (F_{290} - 1)T \tag{5.44}$$

or

$$F_{290} = 1 + \frac{T_e}{T} \tag{5.45}$$

with $T = 290$ K as the reference temperature.

To illustrate the use of these equations consider a satellite receiver whose noise figure is 4. Expressed in decibels this is 6 dB. The effective input noise temperature is thus

$$T_e = (4 - 1) \times 290 = 870 \text{ K}$$

or

$$T_e = 10 \log 870 = 29.4 \text{ dB}.$$

The relationship between noise figure F(dB) and effective noise temperature T_e(K) can be readily obtained from the nomogram in Fig. 47. This is usually useful when considering the G/T ratio.

5.5.3 Total system noise temperature

The two parameters which primarily define the capability of a system to receive and demodulate information in the presence of noise were found to be the overall gain of the system and its noise temperature or noise figure. Attention is here focused on the total system noise temperature. A simplified block diagram of a satellite receiving station is shown in Fig. 48. For the purpose of a generalised gain and noise analysis this may be redrawn as shown in Fig. 49. As indicated earlier, the input of the receiver is used as reference point for

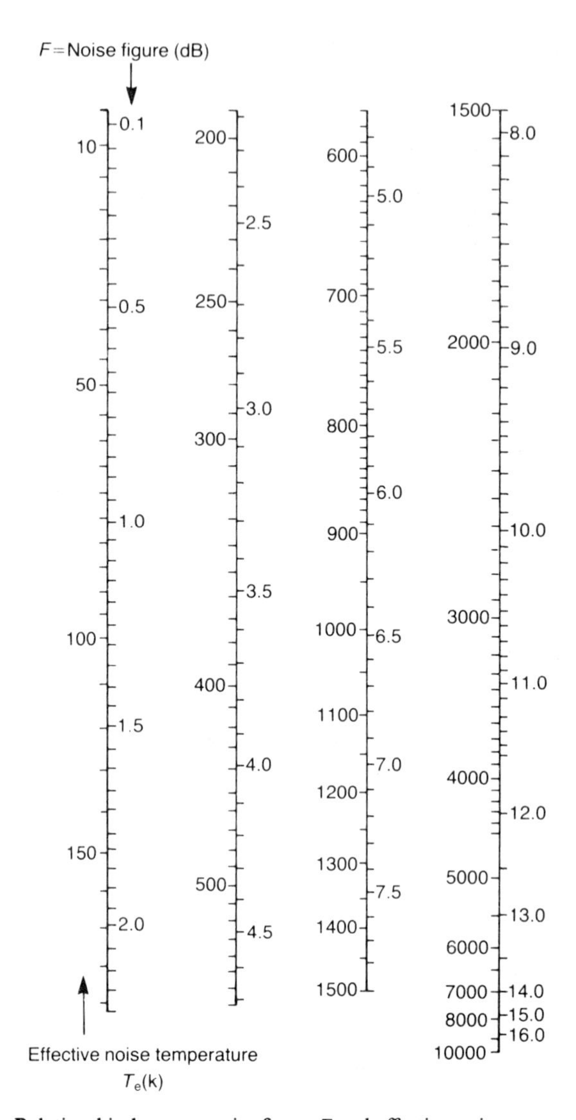

Fig. 47 – Relationship between noise figure F and effective noise temperature T_e.

calculations. The total system noise temperature is then given by

$$T_{e_{sys}} = T_{e1} + \frac{T_{e2}}{G_1} + \frac{T_{e3}}{G_1 G_2} + \ldots . \tag{5.46}$$

In this equation all noise temperatures are expressed in K and all gains as numerical ratios. For devices which are specified by gain and noise figure we

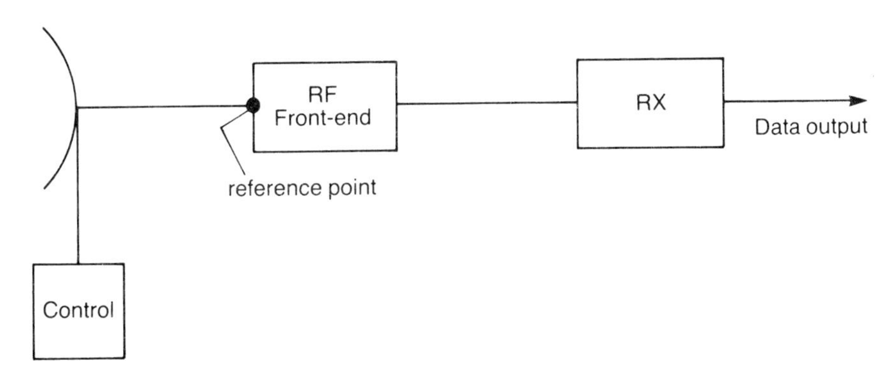

Fig. 48 – Simplified block diagram of a satellite receiving station.

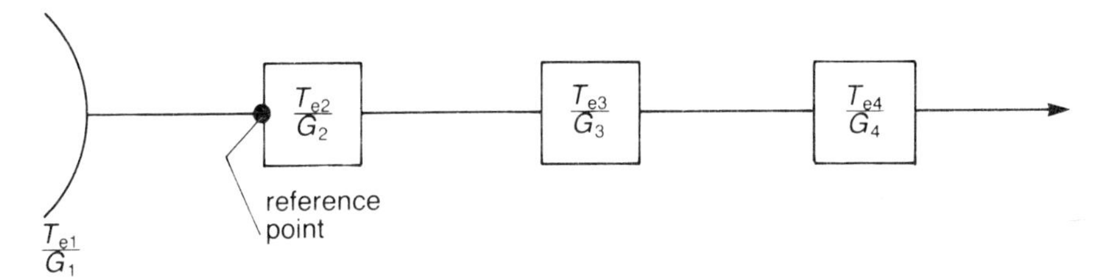

Fig. 49 – Generalized block diagram for system gain and noise analysis.

obtain T_e from

$$T_e = (F - 1)T, \tag{5.47}$$

and for devices which exhibit a loss L and for which no noise figure is specified T_e is given by

$$T_e = (L - 1)T. \tag{5.48}$$

The effective noise temperature for a 6 dB attenuator is thus 290 K (3.98 – 1) = 864.2 K.

A graphical presentation of the elements which contribute losses and noise to a satellite receiver is given in Fig. 50.

As a second example, consider a low-noise amplifier as shown in Fig. 51 directly connected to an aerial of 20 dB gain and a noise temperature of 100 K. What are the equivalent input noise temperature of the amplifier, the total equivalent input noise temperature, and the overall noise figure? Compare the results obtained for noise figure and noise temperature.

From equation (5.44) we obtain for the amplifier equivalent input noise temperature

$$T_{e2} = (F_{290} - 1)290 = (2 - 1)290 = 290\,\mathrm{K}. \tag{5.46}$$

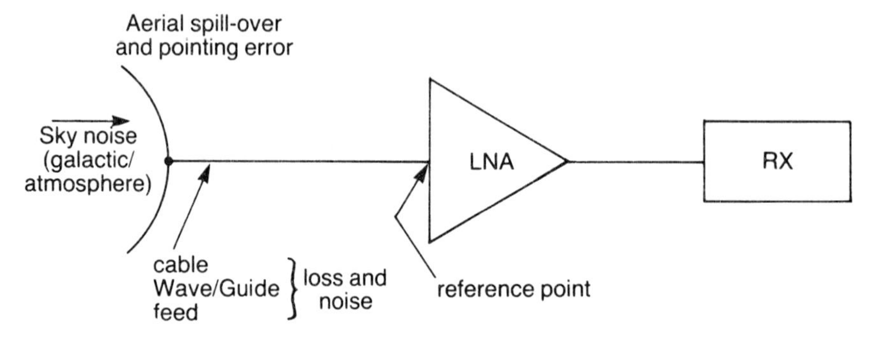

Fig. 50 – Graphical representation of loss and noise contributing elements.

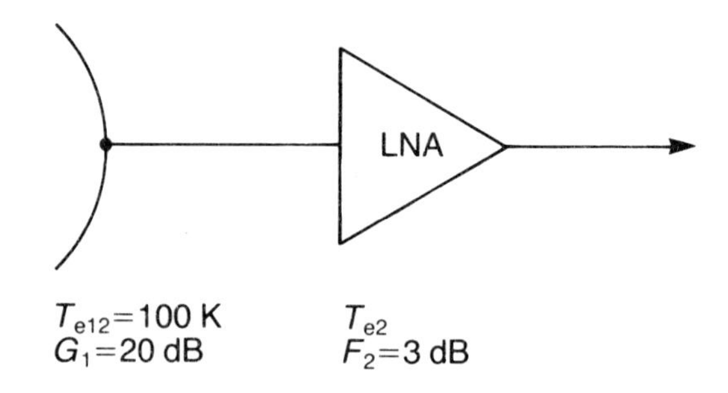

Fig. 51 – Aerial with low-noise amplifier.

The total equivalent input noise temperature is, from equation (5.46)

$$T_{e_{sys}} = T_{e1} + \frac{T_{e2}}{G_1} = 100 + \frac{290}{100} = 100.29.$$

The overall noise figure is obtained from equations (5.38) and (5.45) as

$$F = F_1 + \frac{F_2 - 1}{G_1} = \left(1 + \frac{T_{e1}}{\mathrm{T}}\right) + \frac{F_2 - 1}{G_1}$$

$$F = \left(1 + \frac{100}{290}\right) + \frac{2-1}{100} = 1.345 \text{ or } 1.28 \text{ dB}.$$

This noise figure may be checked against the noise temperature

$$T_{e_{sys}} = (F - 1)290 = (1.345 - 1)290 = 100.05.$$

For all practical purposes both methods provide the same result.

5.5.4 G/T ratio

A receiving system has usually a specification attached to it which defines its ability to correctly receive data in the presence of noise. This ability is frequently referred to as the 'sensitivity' of the system. The sensitivity states the lowest received signal level for which the system will work without, for example, exceeding a desired error rate or distortion. The sensitivity specification is also referred to as the 'threshold' of the system (see also threshold in FM and threshold extension). The threshold can be described by two factors, the receiver aerial gain G and the total system noise temperature T of the receiving system. When the two factors are combined into a single parameter then this is known as the G/T ratio, which is a figure of merit. The usefulness of the G/T ratio lies in the fact that it does not restrict the system designer in the choice of possible solutions. For example, a desired G/T ratio can be obtained either by a small aerial and a low-noise receiver or by a large aerial and a receiver with medium noise figure. If the minimum aerial gain and the maximum receiver gain were specified, then the above choice would not exist. Before calculating the G/T ratio the designer must decide on the minimum antenna elevation and minimum safety margin which he wishes to support. The G/T ratio referred to the input of the receiver can be expressed as

$$\frac{G}{T} = \frac{\alpha\beta G}{\alpha T_a + (1-\alpha)T_o + (F-1)T_o} \qquad (5.49)$$

where T_o = absolute temperature, 290 K
T_a = effective aerial temperature
G = aerial gain
F = receiver noise figure
α = coupling losses
β = pointing and polarisation losses.

For $T_a = 200$ K, $G = 20\,000$, $F = 8$ dB, $\alpha = 1$ dB, $\beta = 0.5$ dB, a G/T ratio of about 4.5 dB/K is obtained.

Problem 7

Calculate the noise figure of the circuit configuration shown in Fig. 45.

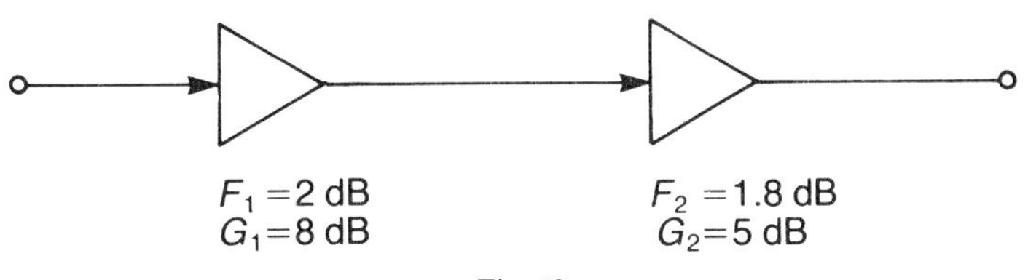

Fig. 52.

Answer

$$F\,[\mathrm{dB}] = 10 \log F \qquad F = \mathrm{alog}\frac{F\,[\mathrm{dB}]}{10}$$

$$F_1 = \mathrm{alog}\frac{2}{10} = 1.58 \quad F_2 = \mathrm{alog}\frac{1.8}{10} = 1.51$$

$$G_1 = \mathrm{alog}\frac{8}{10} = 6.30 \quad G_2 = \mathrm{alog}\frac{5}{10} = 3.16$$

$$F = 1.58 + \frac{1.51 - 1}{6.30} = 1.66 \equiv 2.20\ \mathbf{dB};$$

now interchange amplifiers:

$$F = 1.51 + \frac{1.58 - 1}{3.16} = 1.693 \equiv 2.28\ \mathbf{dB}.$$

Problem 8

A satellite down-converter consisting of a low-noise amplifier, mixer, local oscillator, and intermediate amplifier is shown in Fig. 53. Calculate the overall noise figure of the converter.

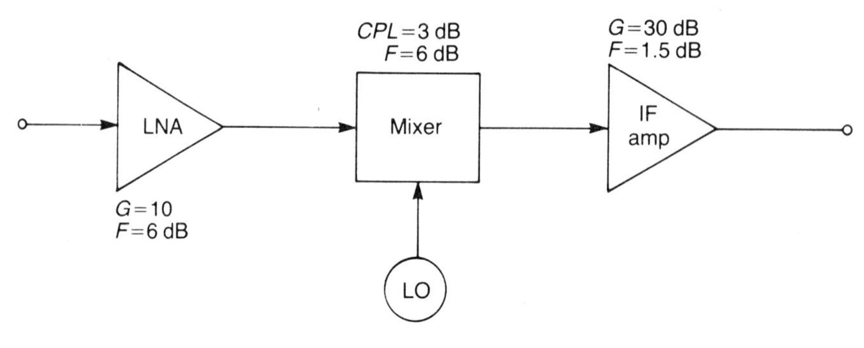

Fig. 53.

Answer

$$F_1 = \mathrm{alog}\frac{3}{10} = 1.995$$

$$F_2 = \mathrm{alog}\frac{6}{10} = 3.981$$

$$F_3 = \mathrm{alog}\frac{1.5}{10} = 1.412$$

$$G_1 = 10$$

$$G_2 = \frac{1}{CPL} = 1\bigg/\left(\text{alog}\frac{3}{10}\right) = 0.5$$

G_3 does not enter calculation

$$F_{\text{tot}} = F_1 + \frac{F_2 - 1}{G_1} + \frac{F_3 - 1}{G_1 G_2}$$

$$= 1.995 + \frac{3.981 - 1}{10} + \frac{1.412 - 1}{10 \times 0.5}$$

$$= 1.995 + 0.298 + 0.0824 = 2.3754$$

$$F_{\text{tot}} = 3.757 \text{ dB}.$$

Problem 9

A 50 Ω parabolic aerial receives a 20 pW signal from a satellite transmission. The aerial is connected via a waveguide, having a loss of 0.5 dB, to a low-noise amplifier of 18 dB gain and 20 K noise temperature. This is followed by a main amplifier having 25 dB gain and a noise figure of 4 dB. The output is connected to a receiver with 8 dB noise figure. Assume an aerial noise temperature of 100 K. Hence

a) draw the block diagram,
b) calculate the total equivalent input noise temperature referred to the waveguide input, and
c) calculate the SNR at the output of the main amplifier for a system bandwidth of 27 MHz.

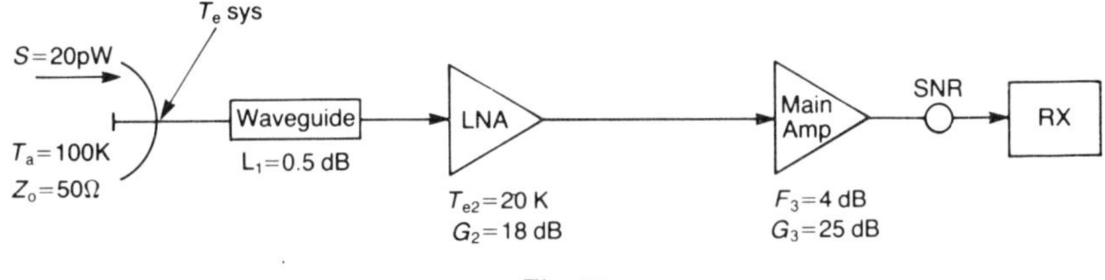

Fig. 54.

Answer

$$L_1 = 0.5 \text{ dB} \equiv 1.122$$

$$T_{e1} = (L_1 - 1)290 = (1.122 - 1)290 = 35.38 \text{ K}$$

$$G_2 = 18 \text{ dB} \equiv 63.09 \qquad F_3 = 4 \text{ dB} \equiv 2.511$$

$$T_{e3} = (F_3 - 1)290 = 438.19 \text{ K}.$$

The total equivalent input noise temperature, also known as system noise temperature, is, from equation (5.46),

$$T_{e\,sys} = T_{e1} + \frac{T_{e2}}{G_1} + \frac{T_{e3}}{G_1 G_2} = 35.38 + \frac{20}{1/1.122} + \frac{438.19}{(1/1.122 \times 63.09)}$$

$$= 35.38 + 22.44 + 7.79 = 65.6 \text{ K}.$$

The total equivalent mean square noise voltage at this point is

$$\overline{e_N^2} = 4kRB(T_{e\,sys} + T_a)$$

$$= 4 \times 1.38 \times 10^{-23} \times 50 \times 27 \times 10^6 (65.6 + 100)$$

$$= 1.23 \times 10^{-11}\ V^2.$$

The signal voltage is

$$e_s^2 = S \times R = 20 \times 10^{-12} \times 50 = 1 \times 10^{-9}\ V^2.$$

The SNR at the main amplifier output is thus

$$\text{SNR} = \frac{e_s^2}{\overline{e_N^2}} = \frac{S}{N} = \frac{10^{-9}}{1.23 \times 10^{-11}} = 81.3 \equiv 19.1 \text{ dB}.$$

Problem 10

Calculate the G/T ratio for a community 12 GHz TVRO of 7 dB noise figure employing a 3 m parabolic dish of 50% efficiency at a reference temperature of 290 K. Assume 0.5 dB coupling and pointing losses each, and an aerial temperature of 180 K. Make use of the nomogram of Fig. 65.

Answer

$$\frac{G}{T} = \frac{0.5 \times 0.5 \times 89\,125}{0.5 \times 180 + 0.5 \times 290 + 6 \times 290} = 11.28 \frac{\text{dB}}{\text{K}}.$$

5.6 RECEIVING AERIALS FOR DBS

A great deal of thought has gone into the best type of aerial [4, 5, 6, 7] for domestic use of 12 GHz, and although many alternatives have been examined it seems that the only practicable design at the present time is the parabolic dish aerial, since this gives the highest gain for the lowest amount of money.

Various designs of flat panel aerial have been examined in attempts to come up with aerials that are less obtrusive than dishes, and these include phased arrays of what are effectively multiple slot aerials cut into sheets of aluminium, so that the finished product looks rather like a sheet of expanded metal. Much research is currently going on into the development of this type of aerial, but at the present time the gain that can be achieved is rather low. Large arrays of printed-circuit dipoles could theoretically be coupled together and appropriately phased to provide a satisfactory beam pattern and a reasonable amount of gain, but although the manufacturing techniques for this type of assembly are well understood, no such designs have yet appeared. Large numbers of semiconductor elements could theoretically be assembled together on flat panels to give the required gain, but this approach is likely to prove expensive, and it is felt that such designs are many years away from being realised.

We showed earlier that the calculations for the WARC plan were carried out on the basis that it would be reasonable for viewers to use parabolic dishes of 0.9 metres diameter, so we shall concentrate on this type of aerial, although it is worth noting that with modern low-noise front ends in receivers, many viewers will probably choose to use smaller dishes because they will be less of an environmental eyesore. Such viewers will probably obtain satisfactory reception for much of the time, but their picture quality will deteriorate noticeably during periods of heavy rain or snow.

The received beam from the satellite can be considered to be a wavefront of parallel rays since it originates tens of thousands of kilometres away. It is a basic property of a parabolic aerial that parallel rays impinging on it are brought to a focus at a point shown as F in Fig. 55. The parabolic shape

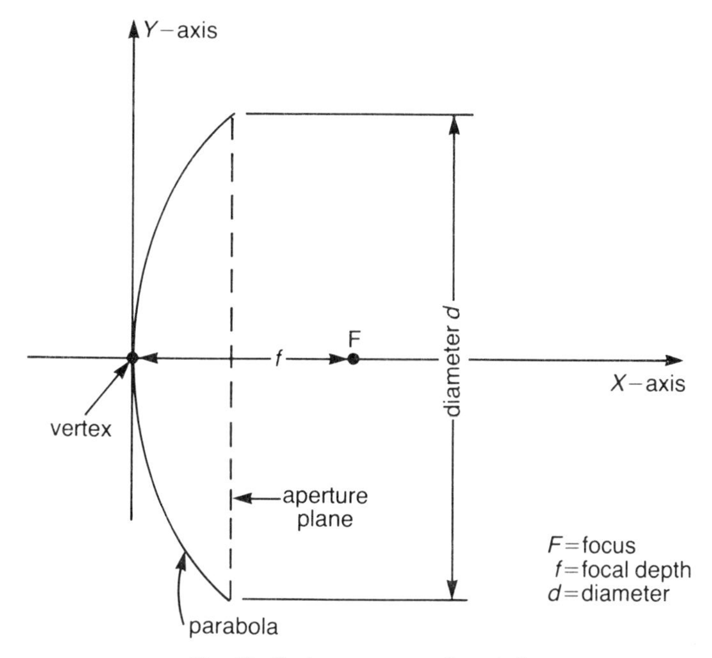

Fig. 55 – Basic parameters of parabola.

effectively converts the parallel wavefront into a spherical wavefront at this point.

The basic parameters of a parabola as used in the context of a receiving aerial are given in Fig. 55. A parabola satisfies the equation $y^2 = 4fx$ as shown in Fig. 56. Considering an incoming plane wave front at the aperture plane, then the path length a b F must be the same for c d F. Sometimes F lies in the aperture plane and the parabolic dish is then referred to as a 'focal plane dish'. The focus-to-diameter ratio (f/d ratio) of practical dishes lies between 0.25 and 0.5. Three examples are given in Fig. 57 together with the most appropriate feed.

There are various ways in which feeds can be mounted in the focus of a dish. In Fig. 58 a length of waveguide is used to hold the feedhorn in position and to carry the signals from the focal point of the dish to the low-noise amplifier. Apart from mechanical considerations, the main drawback of such an arrangement is the considerable signal loss in the waveguide prior to signal amplification. An alternative method is to mount the low-noise amplifier or possibly the complete first down-converter stage directly onto the feed. This, however, has the disadvantage that it will increase the feed weight considerably and will also reduce the effective area of the antenna, which is illuminated by the satellite signal, owing to aperture blockage. Furthermore, the aerial feed has to be dismantled whenever the electronics directly connected to it goes

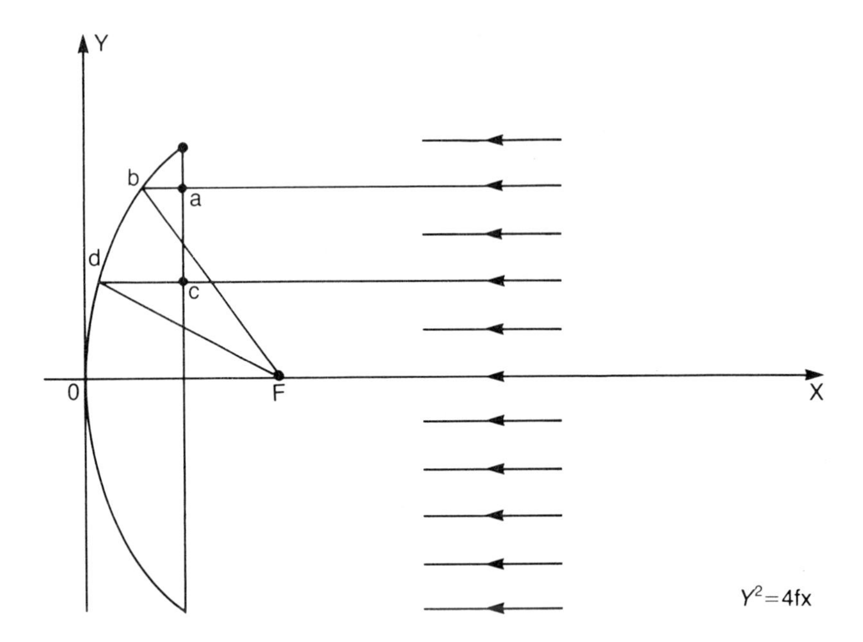

Fig. 56 – Focusing of a planar wavefront.

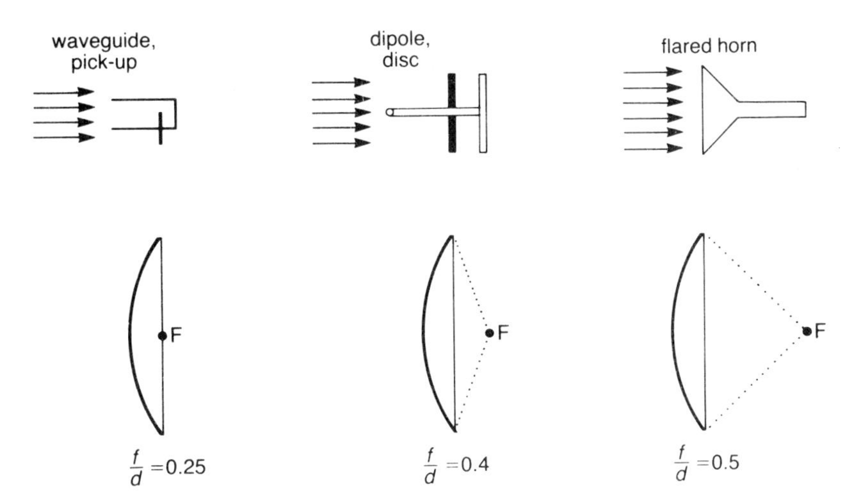

Fig. 57 – Aerial f/d ratio and feeds.

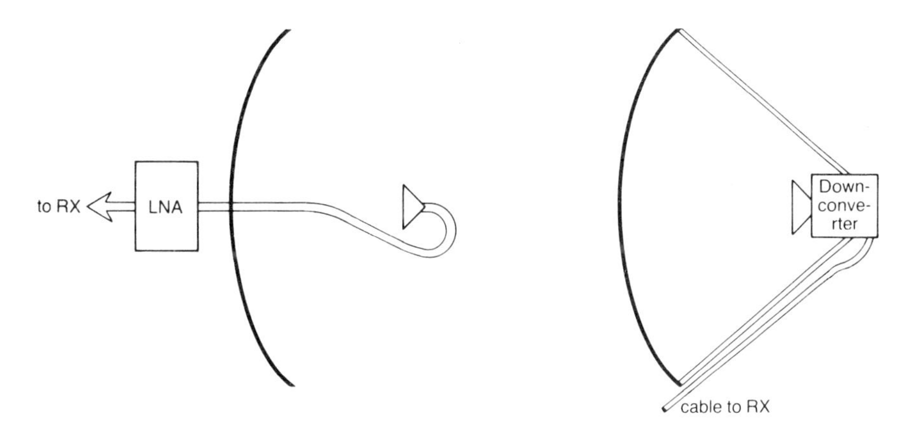

Fig. 58 – Practical aerial feed arrangements.

faulty or needs service. The ultimate solution will depend on the specific satellite link in question. Satellite aerials catering for the general and special needs of customers are manufactured by, for example, Precision Metal Ltd, Masons Road, Stratford-upon-Avon, Warwickshire, England.

Although all the foregoing remarks apply equally well to dishes intended for both up and down links to the satellite, the design of up-link dishes is particularly important because such dishes will be pushing out effective radiated powers of several megawatts, and any side-lobe radiation will be of significant power. Provided that satellites are well separated around the orbit, as in the case of the presently planned direct broadcast satellites which are generally six degrees apart, the specification of standard parabolic dish aerials is usually good enough to keep side-lobe radiation down to levels where interference to adjacent satellites is of sufficiently low level to be unimportant. As time goes on, however, the geostationary orbit is filling up, not only with DBS satellites, which presently comprise only a minority of the orbit's occupants, but also with the ever-growing number of communications satellites and distribution satellites carrying telephone calls, computer data, and television programmes that are the life blood of the information technology revolution. The satellite planners are therefore intending to create more useful space in the orbit by arranging to reduce the satellite separation, perhaps to as little as two degrees. Such developments will call for much more stringent aerial design requirements to prevent interference to satellites in adjacent slots, and it will be very important to keep side-lobe radiation to a minimum.

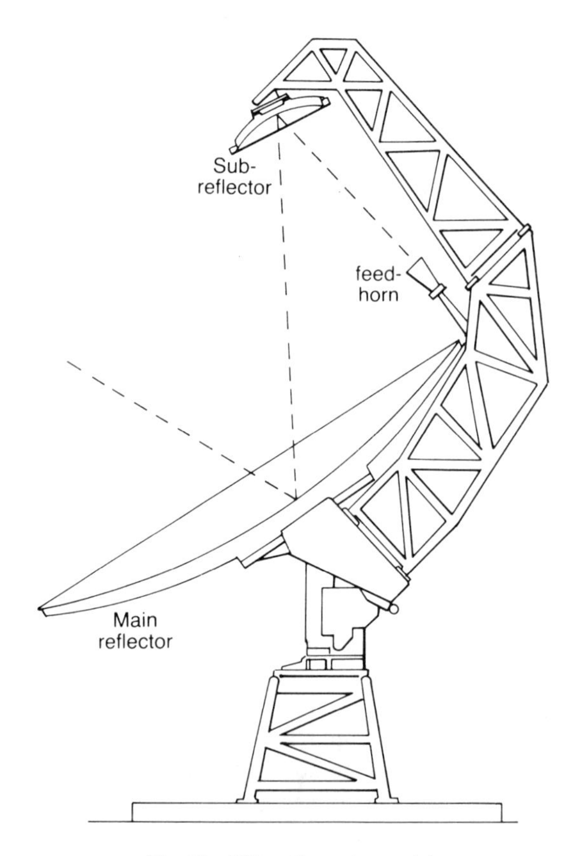

Fig. 59 – Offset Gregorian aerial.

One method of improving the performance of a parabolic antenna system is to use a carefully designed offset-feed arrangement which has been optimised for the particular application. Fairly typical of this new breed of aerial is the so-called offset Gregorian design which British Telecom will be using to feed signals to Intelsat V as part of its 'Satstream North America' service. As Fig. 59 shows, above the 5.5 metre diameter main dish there is a sub-reflector mounted on a lattice-work tower some four metres high, whilst the actual feed horn is offset to the side of the dish. Such a design also has the advantage that virtually the whole of the main dish area is unobstructed, leading to greater efficiency, and the dish itself is nearer to the horizontal than conventional designs for the same effective angle of elevation. The makers claim that this feature will make their dish less obtrusive, although any dish of these dimensions is hardly going to enhance its surrounding area.

Some of the problems indicated earlier can also be overcome, at a price, with the Cassegrain aerial shown in Fig. 60. The electromagnetic energy received by the dish is reflected by means of a small hyperbolic sub-reflector to the feed horn located at the vertex of the dish. The low-noise amplifier can now be mounted directly to the feed without any of the constraints imposed on the designs shown in Fig. 58.

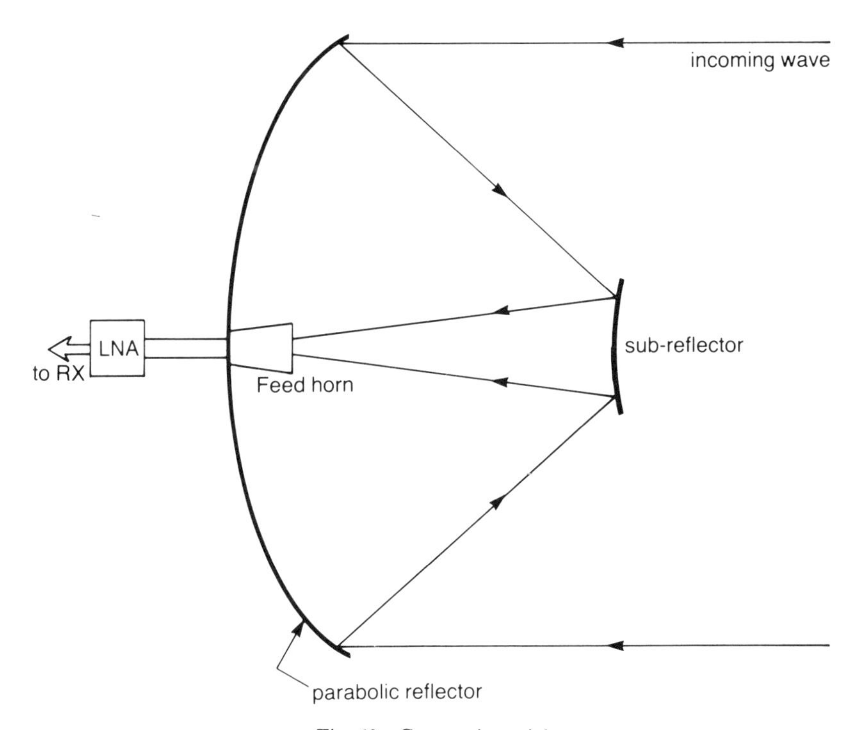

Fig. 60 – Cassegrain aerial.

Summarising, it can be said that the main factors which determine the performance of an aerial, i.e. the attainable gain, are:

a) aperture area,
b) aperture blockage owing to sub-reflector or feed system,
c) aerial profile accuracy,
d) defocusing of sub-reflector and feed system,
e) losses between feed and first low-noise amplifier or mixer,
f) mismatch between free space/feed, feed/waveguide, and waveguide/low-noise amplifier,
g) deformation of dish profile.

After this general discussion on a parabolic dish aerial let us now consider some of the theoretical aspects which permit the gain and radiation pattern of a dish aerial to be calculated.

An important theoretical aerial which radiates equally well in all directions is the point source or isotropic radiator as shown in Fig. 61.

This aerial provides the basis for the comparison of the performance of many different types of aerials. For example, a dish aerial can be made to radiate or receive power from a particular direction; it thus exhibits gain. This is defined as the power gain G

$$G = \frac{\text{power radiated by aerial}}{\text{power radiated by isotropic aerial}} \tag{5.50}$$

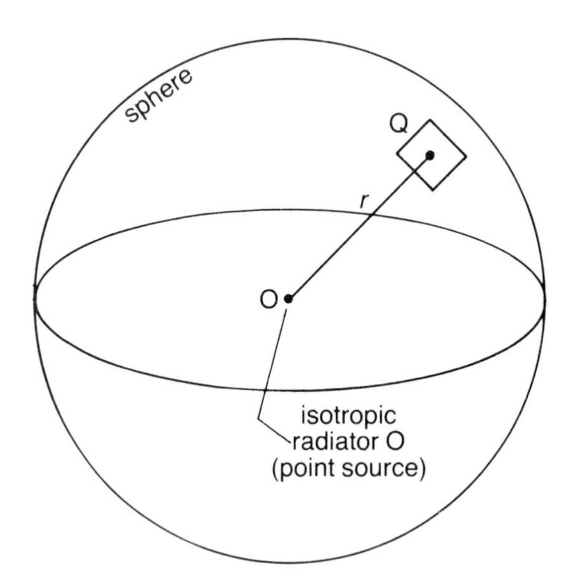

Fig. 61 – Isotropic radiator.

Both aerials are supplied with the same input power. From Fig. 61, the total power transmitted to the spherical surface is

$$P_t = 4\pi r^2 P_r = A P_r \tag{5.51}$$

where P_t = total transmitted power from point source (W),
r = radius of sphere (m),
P_r = power density or Poynting vector (W/m^2),
A = area (m^2).

For an elementary area around Q the power density is therefore only a fraction of the total power. For a parabolic dish the area A is given by

$$A = \pi d^2/4, \tag{5.52}$$

where d is the diameter of the dish. Since the attainable gain for a dish aerial depends on many factors, an efficiency coefficient η is introduced to take account of this. The product ηA is frequently used and is known as 'effective aperture area'.

Based on the definition given in equation (5.50), the power gain G of a dish aerial is given by

$$G = \frac{4\pi\eta A}{\lambda^2} \tag{5.53}$$

where η = aerial efficiency
A = circular area of dish
λ = wavelength received.

Substituting equation (5.52)

$$G = \frac{\pi^2 d^2 \eta}{\lambda^2} \tag{5.54}$$

or with $c = \lambda f$

$$G = \frac{\pi^2 d^2 \eta f^2}{c^2}. \tag{5.55}$$

Apart from the gain the relative radiation pattern or polar diagram of the dish aerial is of interest. This is a plot of power or field strength as a function of aerial angular direction. One distinguishes between horizontal and vertical polar diagrams. In order to understand how a polar diagram may be obtained, consider Fig. 62. The Poynting vector P_r is radial to the imaginary sphere, and a small area around the base of the vector can be looked upon as essentially flat. The electric field E is transverse to the vector, and its two components in the ϕ and θ direction are given by E_ϕ and E_θ. The total electric field is given by $E^2 = E_\phi{}^2 + E_\theta{}^2$.

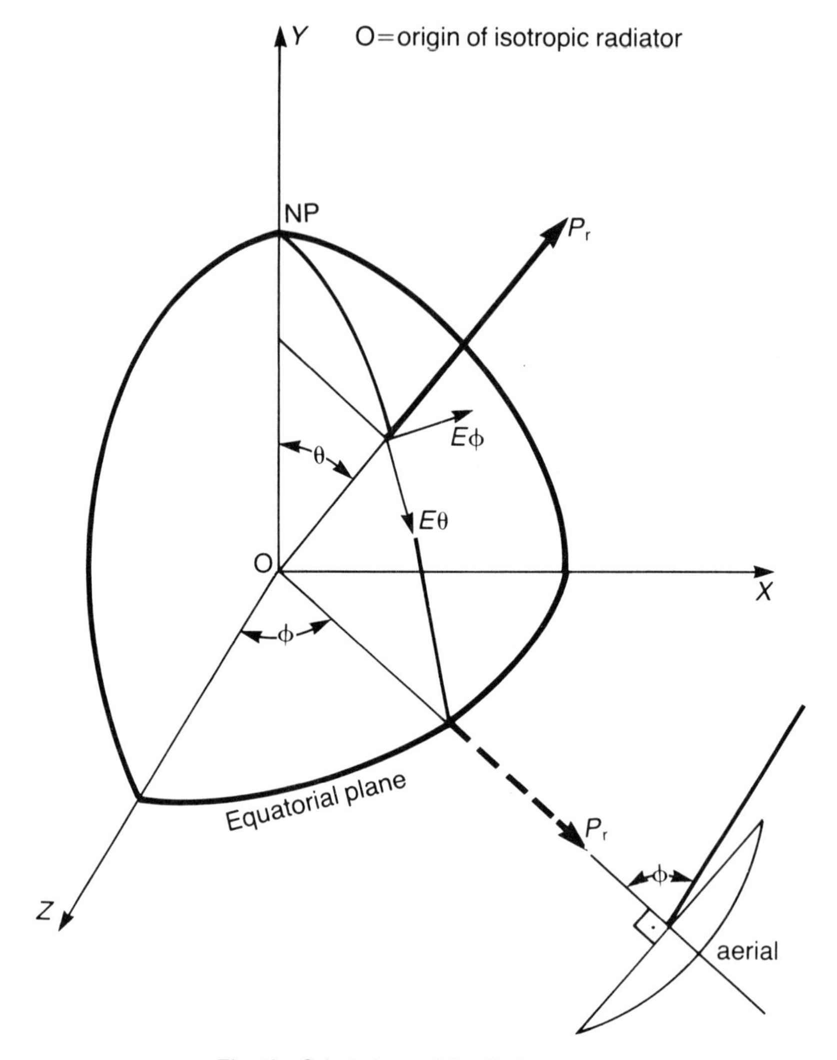

Fig. 62 – Calculating aerial radiation pattern.

The magnetic field H can be obtained from the electric field. For simplicity let the Poynting vector fall into the equatorial plane, i.e. $\theta = 90°$. We are then only left with the component E_ϕ. If the Poynting vector is now rotated about its origin O, then the dish aerial will exhibit the normalised pattern shown in Fig. 63. Employing Huygens' principle, the relative radiation pattern can be calculated from the following equation:

$$E_\phi = \frac{2\lambda J_1\left(\dfrac{\pi d}{\lambda}\sin\phi\right)}{\pi d \sin\phi} \tag{5.56}$$

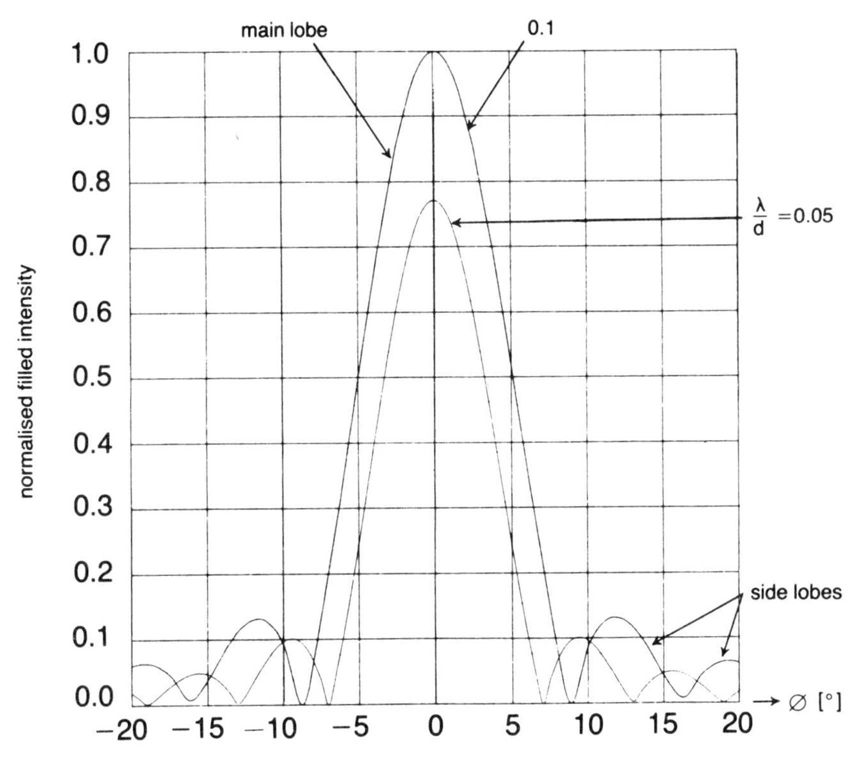

Fig. 63 – Normalised radiation pattern for a parabolic dish with $\lambda/d = 0.1$ and 0.05.

where J_1 = Bessel function of first order,
d = aerial diameter,
λ = wavelength received,
ϕ = angle with respect to normal to aerial aperture.

The values of the argument for which the Bessel function becomes zero can be obtained from Fig. 64, i.e. $J_1(x) = 0$ for $x = 3.8$, 7, 10.3, and 13.5. Hence, the angle of the radiation pattern where the first null occurs is at

$$J_1\left(\frac{\pi d}{\lambda}\sin\phi\right) = J_1(3.8) \tag{5.57}$$

or

$$\phi = \arcsin 3.8\frac{\lambda}{\pi d} = \arcsin 1.22\frac{\lambda}{d}. \tag{5.58}$$

The second null occurs at

$$\phi = \arcsin 7\frac{\lambda}{\pi d} = \arcsin 2.2\frac{\lambda}{d}. \tag{5.59}$$

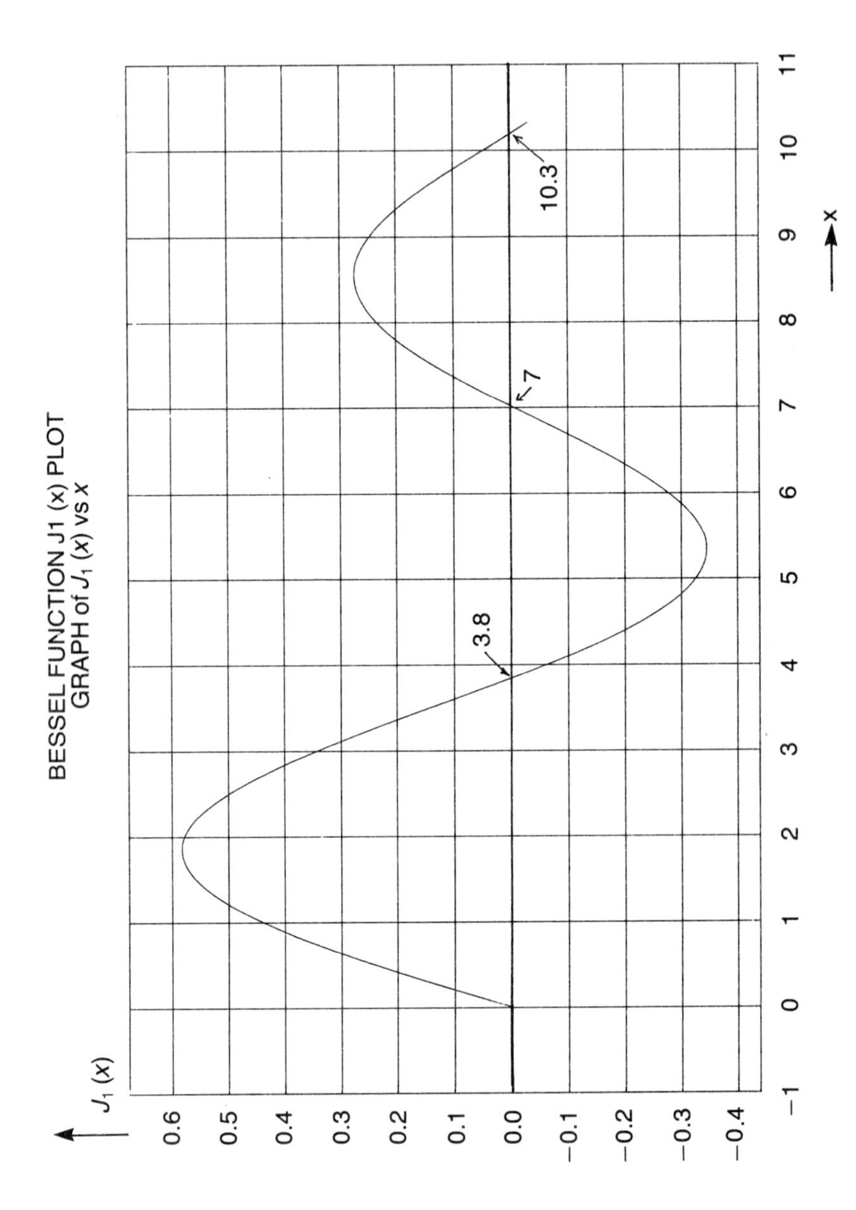

Fig. 64 – Bessel function.

The nulls for subsequent side-lobes are calculated in a similar manner. Of greatest interest is the main lobe of the dish aerial where the gain is a maximum. In the vicinity of maximum gain the angle ϕ is small, and the following approximation can be used to calculate the first null:

$$\phi\,(\text{rad}) = 1.22\frac{\lambda}{d} \tag{5.60}$$

or

$$\phi\,(^\circ) = 70\frac{\lambda}{d} \tag{5.61}$$

The beamwidth of the main lobe between the first nulls is thus twice the value of equation (5.61). The 3 dB or half-power beamwidth is given by

$$\phi_{3\,\text{dB}}\,(^\circ) = 58\frac{\lambda}{d} \tag{5.62}$$

A quick estimate of the aerial parameters can be obtained with the aid of the nomogram shown in Fig. 65. Let us assume one would like to receive with a 3 m parabolic dish aerial the 1.7 GHz information from METEOSAT. What gain can one expect if the aerial has 50% efficiency? The solution is as follows:

1. Draw line through 1.7 GHz frequency scale, parallel to reference line. This line is called 'a'.
2. Draw vertical line to 3 m diameter axis. This line is called 'b'.
3. Lines 'a' and 'b' intersect at A. Draw horizontal line from A to intersect with beamwidth scale to give B.
4. Draw line from $\eta = 50\%$ through B to intersect with gain scale. Read off gain G as approximately 32 dB.

A computer programme which permits the calculation of all relevant parameters of the aerial is given in the Appendix.

Problem 11

A 3 m dish aerial is used for the reception of DBS at 12 GHz. The aerial efficiency is estimated to be 50%. Obtain the aerial gain with the aid of the nomogram.

Answer

Aerial gain = 49.5 dB

(see Fig. 66)

Problem 12

Calculate the efficiency required from a 3 m dish aerial if a signal of 1.7 GHz is to be received from METEOSAT with a gain of 33 dB. Make use of the nomogram in Fig. 65.

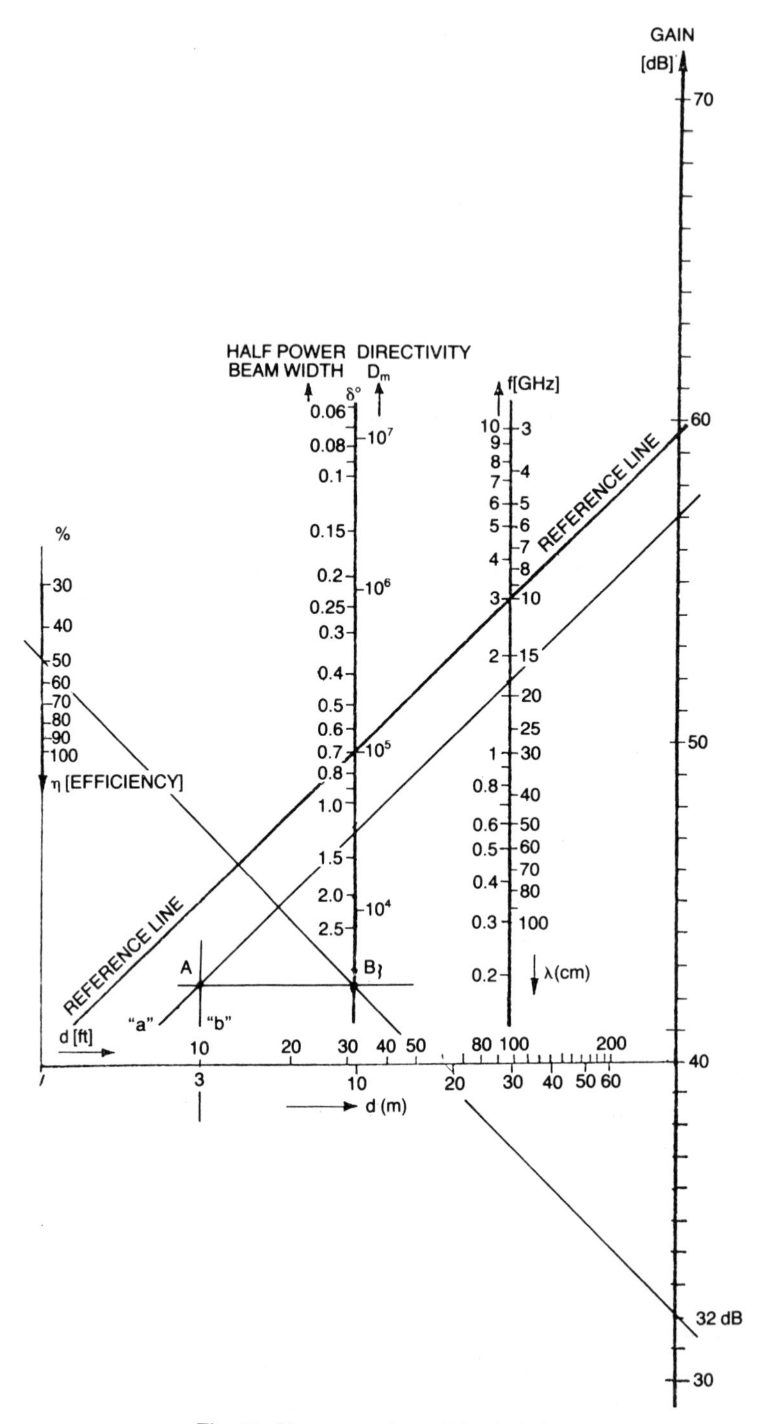

Fig. 65 – Nomogram for aerial calculations.

SOLUTION PROBLEM 8

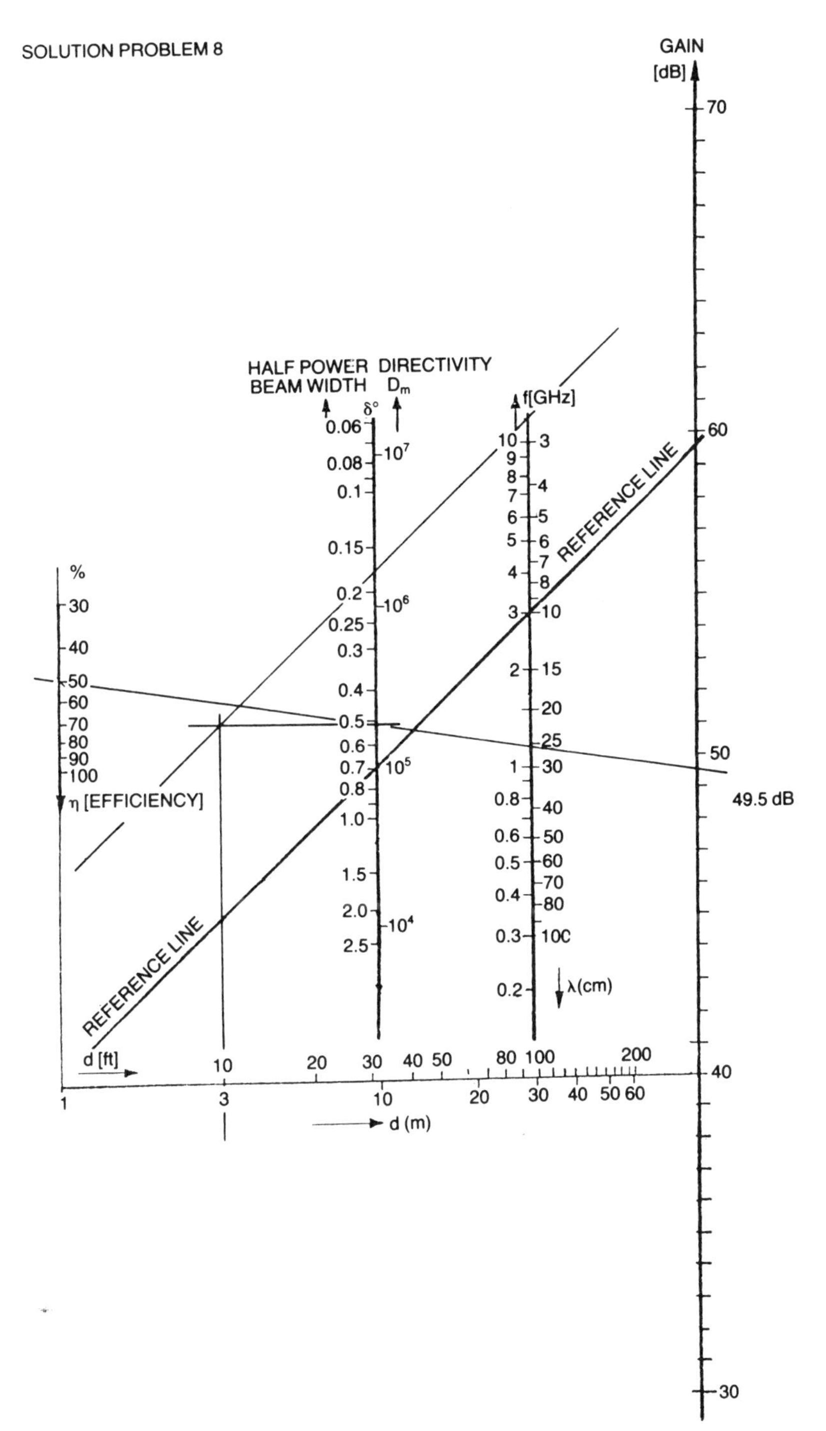

Fig. 66.

SOLUTION PROBLEM 8

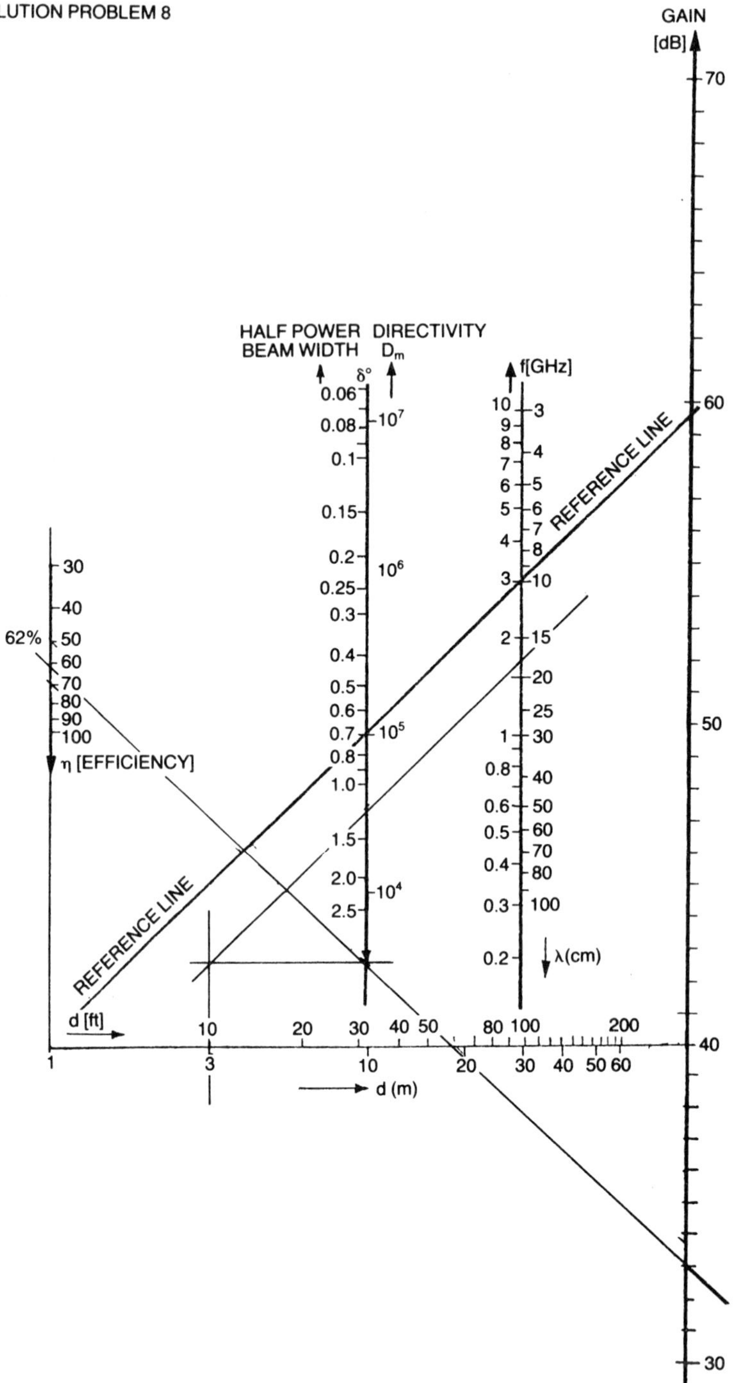

Fig. 67.

Answer

Efficiency = 62 %

See Fig. 67

Problem 13

An aerial working at 1.7 GHz and having a diameter of 10 m is required to have a gain of 43 dB. Obtain the necessary efficiency. How much aerial gain variation is associated with ±5 % efficiency variation?

Answer

$\eta = 63\,\%$; for $\eta = 63\,\% \pm 5\,\%$ we obtain $G = 43\text{ dB} \pm 0.21\text{ dB}$

Problem 14

Verify the normalised radiation pattern diagram up to the third side-lobe as shown in Fig. 63. Assume an aerial diameter of 176.4 cm and an operating frequency of 1.7 GHz.

Answer

$$\lambda = \frac{c}{f} = \frac{3 \times 10^{10}\text{ cm s}}{1.7 \times 10^{9}\text{ s}} = \frac{30}{1.7} = 17.64\text{ cm} \quad \frac{\lambda}{d} = \frac{17.64}{176.4} = 0.1$$

$$\phi_0 = \sin^{-1}\frac{x\lambda}{\pi d} \quad \text{for nulls}$$

From Bessel curves:

$J_1(x)$ is zero for $x = 3.8$, 7, 10.3, and 13.5

First null

$$\phi_{01} = \sin^{-1}\left(\frac{3.8 \times 0.1}{\pi}\right) = 6.94^\circ$$

$$\phi_{02} = \sin^{-1}\left(\frac{7 \times 0.1}{\pi}\right) = 12.87^\circ$$

$$\phi_{03} = \sin^{-1}\left(\frac{10.3 \times 0.1}{\pi}\right) = 19.13^\circ$$

$$\phi_{04} = \sin^{-1}\left(\frac{13.5 \times 0.1}{\pi}\right) = 25.44^\circ.$$

Max. of 1st side lobe $\frac{6.94^\circ + 12.87^\circ}{2} = 9.9^\circ$. Hence using eqn (5.56)

$$E_\phi = \frac{2\lambda J_1\left(\frac{\pi d}{\lambda}\sin\phi\right)}{\pi d \times \sin\phi}$$

$$E_\phi = \frac{2}{\pi} \times 0.1 \times \frac{1}{\sin 9.9^\circ} \times J_1\,(\pi \times 10 \times \sin 9.9^\circ)$$

$$= 0.06366 \times \frac{1}{0.1719} \times \quad J_1\,(5.4) \quad \text{is 0.35 from Bessel curve}$$

$$= \underline{0.13}$$

Maxima of other side lobes can be calculated in a similar manner.

Maximum of 2nd side lobe at $\frac{12.87^\circ + 19.13^\circ}{2} = 16^\circ$ is $\underline{E_\phi = 0.07}$

try 5° $\quad E_\phi = \underline{0.34}$

try 1° $\quad E_\phi = \underline{0.91}$ approx.

Problem 15

Calculate the normalised radiation pattern up to the third side-lobe for an aerial dish of 3.5 m diameter working at a center frequency of 1.7 GHz. Superimpose this diagram on that of the previous problem. At what conclusions do you arrive?

Answer

$$\frac{\lambda}{d} = 0.05 \quad \text{The nulls are at } \phi_0 = \sin^{-1}\frac{x\lambda}{\pi d}$$

From Bessel curves $J_1\,(x)$ is zero for $x = 3.8$, 7, 10.3, and 13.5

$$\phi_{01} = \sin^{-1}\left(\frac{3.8 \times 0.05}{\pi}\right) = 3.46^\circ;\ \phi_{02} = \sin^{-1}\left(\frac{7 \times 0.05}{\pi}\right) = 6.4^\circ$$

$$\phi_{03} = \sin^{-1}\left(\frac{10.3 \times 0.05}{\pi}\right) = 9.43^\circ;\ \phi_{04} = \sin^{-1}\left(\frac{13.5 \times 0.05}{\pi}\right)$$

$$= 12.4^\circ$$

Max. of 1st side lobe $\frac{3.46^\circ + 6.4^\circ}{2} = 4.93^\circ \qquad \sin 4.93^\circ = 0.0859$

2nd side lobe $\frac{6.4^\circ + 9.43^\circ}{2} = 7.9^\circ \qquad \sin 7.9^\circ = 0.1374$

3rd side lobe $\frac{9.43 + 12.4}{2} = 10.9°$ sin 10.9° = 0.189.

The E at those angles is $E_\phi = \dfrac{2\lambda J_1\left(\dfrac{\pi d}{\lambda}\sin\phi\right)}{\pi d \sin\phi}$

$$\frac{2\lambda}{\pi d} = 0.03\,183$$

$$E_{\phi 1} = 0.03183 \times \frac{1}{0.0859} \times \underbrace{J_1(62.83 \times 0.0859)}_{5.4} \rightarrow 0.35 \text{ from Bessel curve}$$

$$= \underline{0.13}$$

$$E_{\phi_2} = 0.03\,183 \times \frac{1}{0.1374} \times J_1(62.83 + 0.1374) = 0.23 \times (0.3)$$

$$= \underline{0.07}$$

$$E_{\phi_3} = 0.03\,183 \times \frac{1}{0.189} \times J_1(62.83 \times 0.189) = 0.168 \times (0.22)$$

$$= \underline{0.037}$$

Conclusion: The larger the aerial, the larger the gain, and the narrower the main lobe.

Problem 16

Verify the result obtained in problem 11 by means of the computer program "AERIAL" listed in the appendix.

Answer

```
SUMMARY OF AERIAL DATA

d cm                    300
eta  o/o                50
f GHz                   12
G                       71061.151
G(dB)                   48.516322
delta 3dB,degr.         0.58333334
lambda cm               2.5
```

REFERENCES

[1] G. R. Mueller and E. R. Spangler, Communication satellites. Wiley and sons, New York, 1964.
[2] R. F. Filiponski and T. I. Muehldorf, Space Communication Technique. Prentice Hall.
[3] W. B. Davenport and W. L. Root, *Random signals and noise.* Mac-Graw Hill, New York, 1958.
[4] R. Rosenberg, Broadcast TV-satellite antenna systems. *Microwave Journal*, **24,** No. 1, Jan. 1981, pp. 51.
[5] J. R. Forrest, Assessing TVRO antennas. *Microwave System News*, **11,** No. 10, Oct. 1981, pp. 77.
[6] J. R. James *et al.*, Microstrip antenna performance is determined by substrate constraints. *Microwave System News*, **12,** No. 8, Aug. 1982.

6

Design aspects of satellite receivers

The present chapter describes in more detail the operation and design of the elements which make up a satellite TV receiver. Some of the information provided is the result of experimentation undertaken in this area. Although the hardware is not described down to the last detail, the data provided, together with the references, can be used as a basis for receiver design.

This chapter is broadly divided into the following groups:

- low-noise amplifier
- mixer
- local oscillator
- IF amplifier
- demodulator.

6.1 MICROWAVE AMPLIFIER

Signal levels from satellite transmissions are very small, and there is thus a need for low-noise amplification [1, 2, 3]. A variety of semiconductor devices may be used such as GaAsFETs, bipolar transistors, Gunn diodes, IMPATT devices, and tunnel diodes. Apart from performance considerations it is usually the cost of the device which plays a dominant role in domestic receiver application. On balance a two-stage GaAsFET amplifier satisfies this criterion. It satisfies also the other criterion, namely low noise, high gain, and linear amplification up to about 30 GHz.

Different design approaches must be employed [4] depending on the amplifier mode of operation which is required: low noise, high gain, high power. A typical block diagram of a two-stage GaAsFET amplifier is shown in Fig. 68.

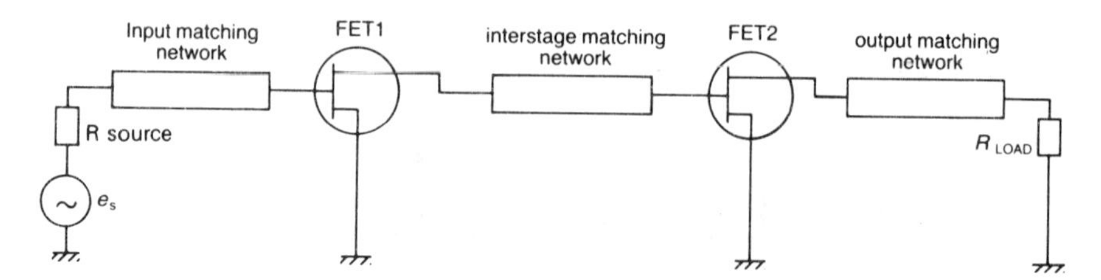

Fig. 68 – Block diagram of a two stage FET amplifier.

The basic difficulty in designing microwave amplifiers lies in calculating the source impedance which must drive the device and the load which the FET must drive in order that at any given frequency the FET will have the required gain. These calculations are non-trivial, since there may be certain values of source impedance and/or load impedance which will cause the amplifier to be unstable. If there are no impedances that cause instability at either the input or output, then the device is said to be unconditionally stable. Unfortunately, however, most FETs are only conditionally stable above about 4 GHz. Once the device has been categorised as either unconditionally stable or conditionally stable at the required frequency, then the design follows two separate paths, i.e. the stability criteria and the matching conditions. Various books and research papers contain detailed treatment of the network theory involved.

The following is a microwave amplifier design which uses a 50 Ω microstrip throughout [5]. The schematic diagram for this type of amplifier is shown in Fig. 69. The capacitors C_1 and C_2 are of such value such that their impedance is small at the operating frequency. To keep the dimensions of the components similar to the width of the 50 Ω line, chip capacitors are used. A possible amplifier layout is shown in Fig. 70.

The $\lambda/4$ line is used to bias the microwave transistors, as will be explained

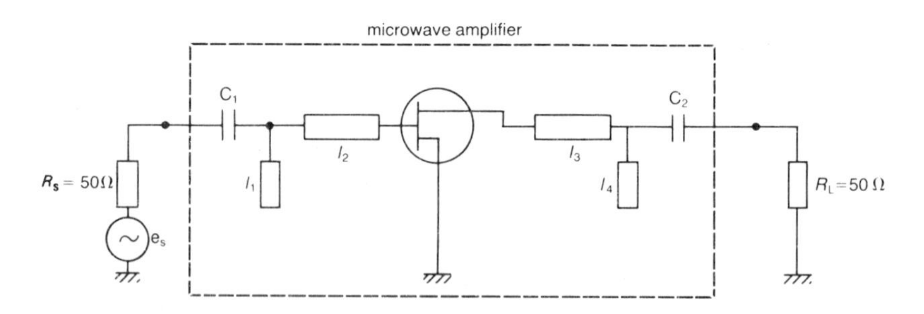

Fig. 69 – Schematic of single-stage FET amplifier using 50 Ω microstrip throughout.

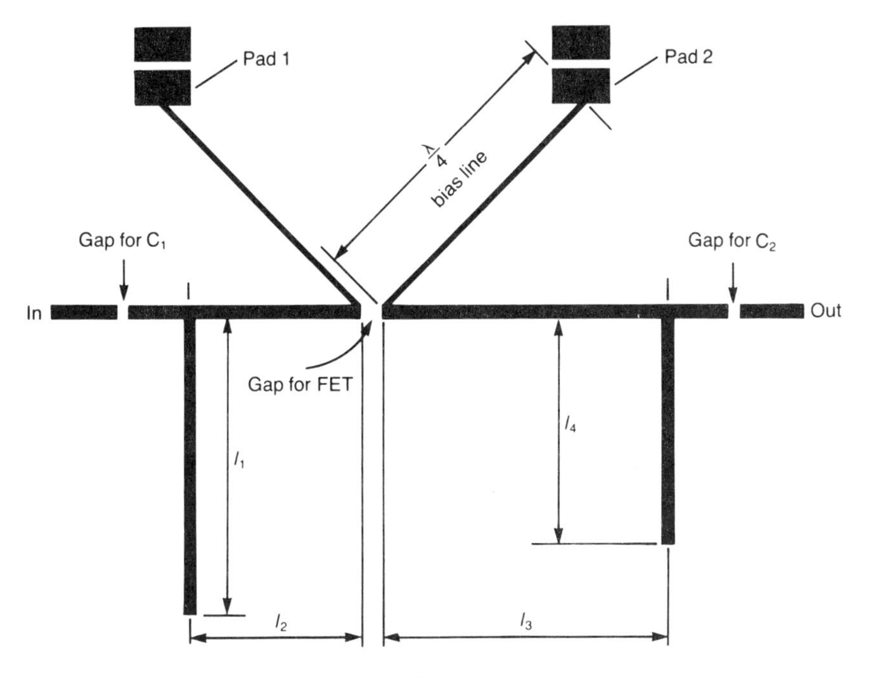

Fig. 70 – A possible amplifier layout.

later. Practical line impedances are 80 to 100 Ω. The microstrip line lengths l_1 to l_4 are known as:

l_1 = input stub
l_2 = input line
l_3 = output line
l_4 = output stub.

The lengths and position of the lines are governed by the transistor parameters, also known as scattering parameters [6], i.e. S_{11}, S_{12}, S_{21}, and S_{22}, where

S_{11} = input reflection coefficient
S_{21} = forward transmission gain
S_{12} = reverse transmission gain
S_{22} = output reflection coefficient.

The calculation of the lengths is quite time-consuming, and it pays to write a computer program if different designs are required. Since the Spectrum computer does not handle complex numbers, a program was written on the Leeds Polytechnic computer. The four scattering parameters are typed in as data, and the computer calculates the four line lengths and, in actual fact, plots the amplifier layout as shown in Fig. 70.

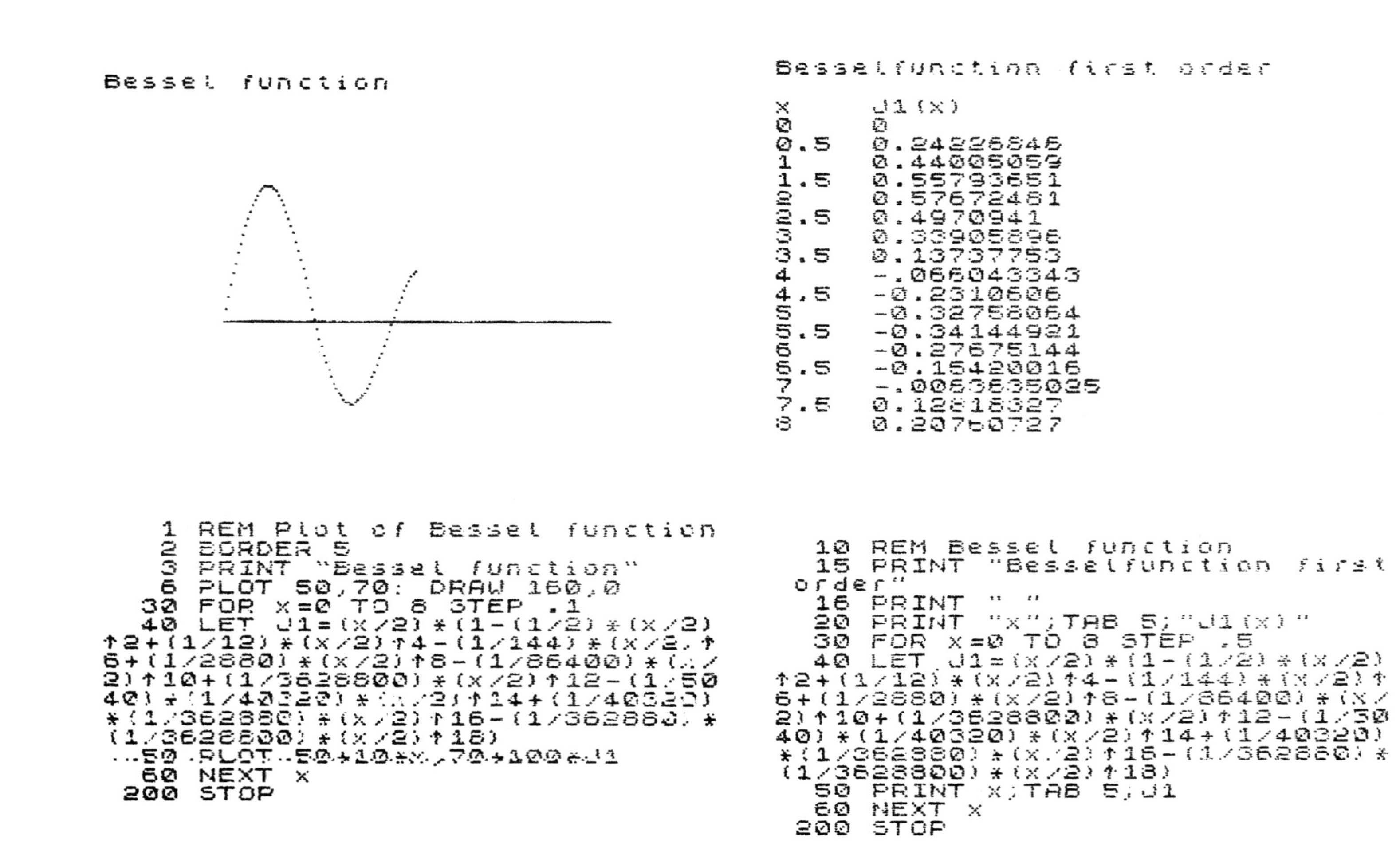
Bessel function

```
  1 REM Plot of Bessel function
  2 BORDER 5
  3 PRINT "Bessel function"
  6 PLOT 50,70: DRAW 160,0
 30 FOR x=0 TO 8 STEP .1
 40 LET J1=(x/2)*(1-(1/2)*(x/2)
↑2+(1/12)*(x/2)↑4-(1/144)*(x/2)↑
6+(1/2880)*(x/2)↑8-(1/86400)*(x/
2)↑10+(1/3628800)*(x/2)↑12-(1/50
40)*(1/40320)*(x/2)↑14+(1/40320)
*(1/362880)*(x/2)↑16-(1/362880)*
(1/3628800)*(x/2)↑18)
 50 PLOT 50+10*x,70+100*J1
 60 NEXT x
200 STOP
```

Besselfunction first order

x	J1(x)
0	0
0.5	0.24226846
1	0.44005059
1.5	0.55793651
2	0.57672481
2.5	0.4970941
3	0.33905896
3.5	0.13737753
4	-.066043343
4.5	-0.2310606
5	-0.32758064
5.5	-0.34144921
6	-0.27675144
6.5	-0.15420016
7	-.0063635025
7.5	0.12818327
8	0.20760727

```
 10 REM Bessel function
 15 PRINT "Besselfunction first
order"
 16 PRINT " "
 20 PRINT "x";TAB 5;"J1(x)"
 30 FOR x=0 TO 8 STEP .5
 40 LET J1=(x/2)*(1-(1/2)*(x/2)
↑2+(1/12)*(x/2)↑4-(1/144)*(x/2)↑
6+(1/2880)*(x/2)↑8-(1/86400)*(x/
2)↑10+(1/3628800)*(x/2)↑12-(1/50
40)*(1/40320)*(x/2)↑14+(1/40320)
*(1/362880)*(x/2)↑16-(1/362880)*
(1/3628800)*(x/2)↑18)
 50 PRINT x;TAB 5;J1
 60 NEXT x
200 STOP
```

```
   1 LPRINT "DISH"
   2 LLIST
  10 REM DISH
  20 REM calculation of gain of
parabolic dish aerial
  30 INPUT "d in cm",d
  40 INPUT "eta in o/o",eta
  50 INPUT "f in GHz",f
  60 LET pi=PI
  70 LET c=3*10↑10
  71 LET lambda=c/(f*10↑9)
  80 LET G=(pi*pi*d*d*(eta/100)*
f*f*(10↑18))/(c*c)
  81 LET delta=73*lambda/d
  89 REM x is gain in dB
  90 LET x=10*LN G/LN 10
  91 PRINT
  92 PRINT
  93 PRINT "SUMMARY OF AERIAL DA
TA"
  94 PRINT
  95 PRINT
 100 PRINT "d cm",d
 101 PRINT
 110 PRINT "eta  o/o",eta
 111 PRINT
 120 PRINT "f GHz",f
 121 PRINT
 130 PRINT "G",G
 131 PRINT
 140 PRINT "G(dB)",x
 141 PRINT
 150 PRINT "delta 3dB,degr.",del
ta
 160 PRINT
 161 PRINT "lambda cm",lambda
 200 STOP
```

Table 6.2 shows the line lengths calculated from the low-noise scattering parameters of the GaAsFET NE72089 given in Table 6.1.

To obtain the actual line length the calculated values e.g. as obtained from Table 6.2, must be multiplied by the effective wavelength, as explained in the section dealing with microstrip lines (Section 7.1).

Table 6.1 GaAsFET S-parameters

S–MAGN AND ANGLES:
$V_{DS} = 3$ V, $I_{DS} = 10$ mA

Frequency (MHz)	S_{11}		S_{21}		S_{12}		S_{22}	
1000	0.98	−31	3.39	154	0.03	73	0.68	−20
2000	0.92	−60	3.25	125	0.06	53	0.66	−39
4000	0.75	−117	2.56	79	0.12	15	0.54	−78
6000	0.67	−160	2.02	43	0.11	−6	0.50	−110
8000	0.62	166	1.64	10	0.12	−21	0.50	−146
10000	0.57	127	1.55	−18	0.14	−38	0.51	178
12000	0.55	86	1.50	−52	0.15	−60	0.55	137

Table 6.2 Line lengths for amplifiers working at given frequency

f (GHz)	l_1 ($\times \lambda_{eff}$)	l_2 ($\times \lambda_{eff}$)	l_3 ($\times \lambda_{eff}$)	l_4 ($\times \lambda_{eff}$)	k	$G_{selected}$ (dB)	MSG (dB)
1	0.223	0.233	0.284	0.391	0.062	19	20.53
2	0.290	0.130	0.476	0.067	0.245	16	17.33
4	0.185	0.143	0.215	0.010	0.526	12	13.29
6	0.313	0.466	0.479	0.385	0.839	11	12.63
8	0.302	0.419	0.473	0.326	0.990	10	11.35
10	0.325	0.362	0.409	0.373	0.968	8	10.44
12	0.311	0.324	0.365	0.362	0.919	8	10.00

Let us assume that we want to receive a 12 GHz satellite transmission and require a single-stage amplifier. How do we go about the design of the amplifier?

First we choose a suitable substrate, remembering the details provided in the microstrip section. Hence

$$Z_o = 50\ \Omega$$
$$\varepsilon_r = 10.5$$
$$h = 1.27\ \text{mm}$$
$$f = 12\ \text{GHz.}$$

Using the microstrip program we obtain the following results:

$$w = 1.159\ \text{mm}$$
$$\varepsilon_{eff} = 7.012$$
$$\lambda_{eff} = 9.44\ \text{mm}$$
$$\lambda_{eff/4} = 2.36\ \text{mm} \qquad = \text{bias line}$$

The actual line length are as follows, using Table 6.2.

$$l_1 = 0.311 \times 9.44 = 2.93\ \text{mm}$$
$$l_2 = 0.324 \times 9.44 = 3.05\ \text{mm}$$
$$l_3 = 0.365 \times 9.44 = 3.44\ \text{mm}$$
$$l_4 = 0.362 \times 9.44 = 3.41\ \text{mm.}$$

The bias lines are 2.36 mm each. The microstrip length to the left and right of the input and output stub can be of any length, but must have an impedance of 50 Ω. The gaps are about one mm, and are wide enough to accommodate the chip capacitor. Two further chip capacitors of about 1000 pF are connected to the pads at the end of the bias line. The uppermost pad is then shimmed with a piece of copper foil to ground as shown in Fig. 71. Alternatively, a hole can be drilled into the bias line pad and the chip capacitor inserted end on and then soldered into place.

The scattering parameters of a transistor are a function of the d.c. bias

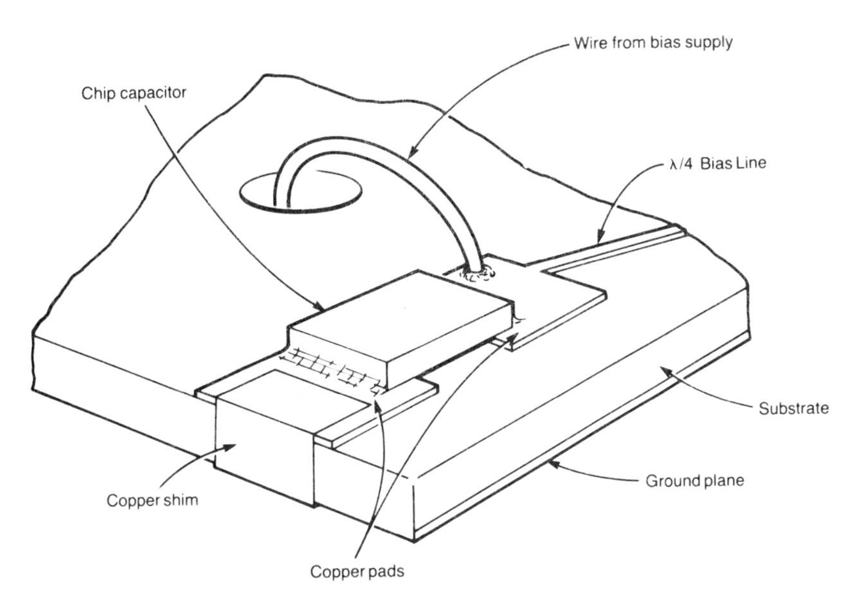

Fig. 71 – Physical arrangement for applying bias.

conditions. A means has to be found for maintaining both the d.c. current through the device and the d.c. voltage across it, that is I_c and V_{CE} for a bipolar transistor or I_{DS} and V_{DS} for a FET. A suitable circuit is shown in Fig. 72. For a single stage amplifier V_{DS1} and V_{GS1} are connected to the drain and gate of the FET and the second stage is omitted. More commonly two stage amplifiers are used in DBS receivers with the first stage current adjusted to give minimum noise figure (say 10 mA) and with the second stage adjusted for maximum gain (say 30 mA).

6.2 MIXERS

Mixers are employed in almost all communication circuits, including the microwave front end of a DBS receiver. Mixers are used for the down conversion of the signal frequency (RF) to a convenient intermediate frequency (IF) for ease of processing or amplification [7, 8]. A mixer consists of a network which contains one or more non-linear devices such as semiconductor diodes. Furthermore, it requires a means of coupling-in the signal and local oscillator (LO) and of coupling-out the IF. A low receiver noise temperature is required in order to detect weak signals. This is to a considerable extent dependant on the mixer noise figure (F) and conversion power loss (CPL).

The simplest mixer is the single-ended mixer. One of its main disadvantages is that there is no isolation between the signal and local oscillator.

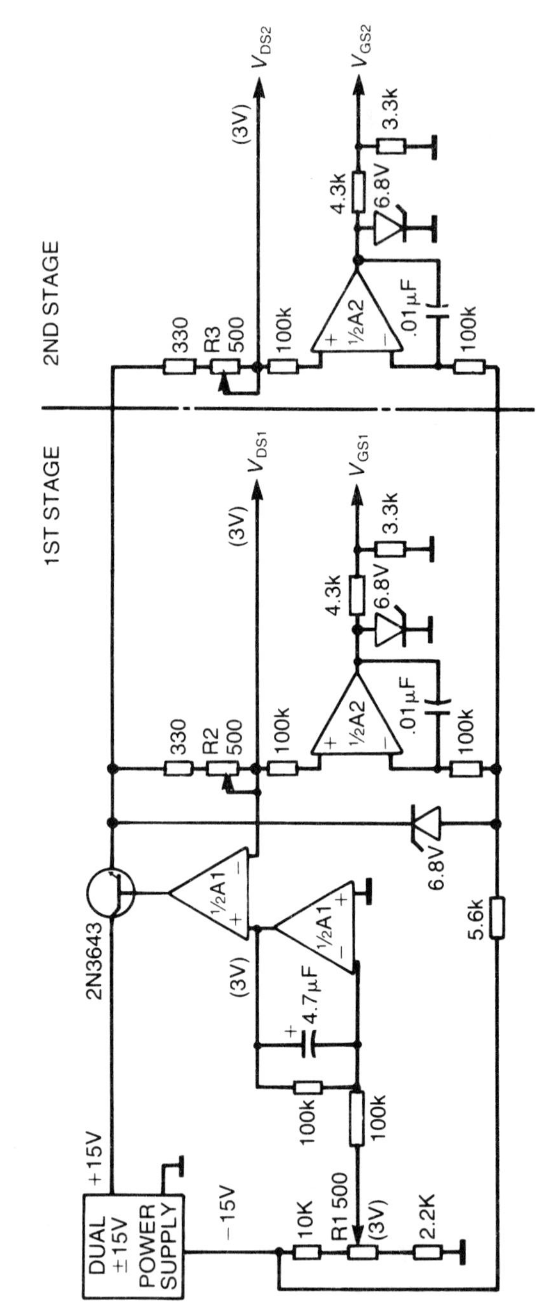

Fig. 72 – One/two-stage FET bias supply.

That means that the local oscillator can propagate out of the signal port. A filter or isolator is then required which in turn reduces circuit simplicity.

A frequently used mixer configuration is the single balanced mixer [9, 10]. A version employing discrete elements is shown in Fig. 73. The elements L_1/C_1 form a high-pass filter and L_2/C_2 a low-pass filter. This allows for the separation of the RF and IF. Diodes D_1 and D_2 are driven by the LO via a balanced transformer T. During the positive half of the oscillator cycle both diodes conduct, thus effectively short-circuiting point X to ground. During the negative cycle the diodes are reverse biased, i.e. open circuit. This type of mixer thus works like a shunt mixer as opposed to a series mixer. The switching action by the diodes together with the signal introduces harmonic frequencies. One of those frequencies is the IF which is extracted by means of appropriate filtering [11, 12].

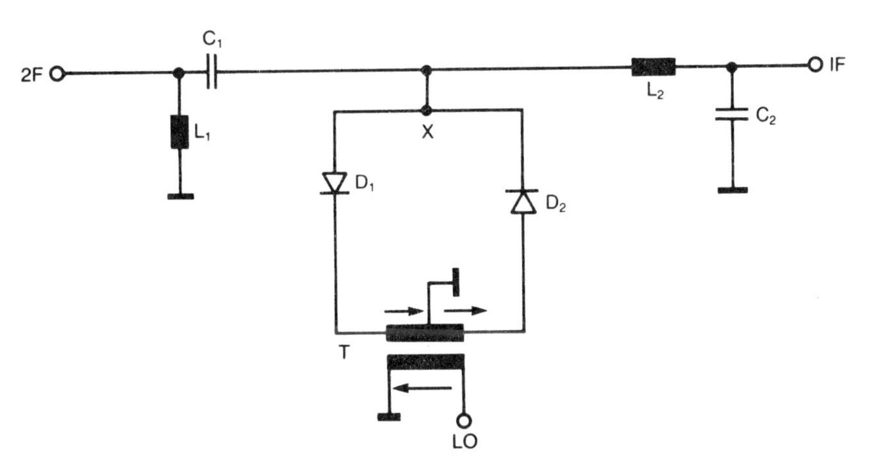

Fig. 73 – Single balanced mixer.

Naturally, lumped elements cannot be used at the higher microwave frequencies, and other means have to be found for circuit construction such as planar transmission line technology. The following are examples of couplers which allow the combining of RF and LO to produce a mixer:

1) hybrid coupler,
2) rat-race,
3) travelling wave directional filter.

The first two are discussed here. Depending on the substrate used, the couplers are suitable for the frequency range of about 1 GHz–40 GHz.

The rat-race, also known as 'ring directional coupler', is shown in Fig. 74. It is circular, is one and a half wavelength in circumference, and has an

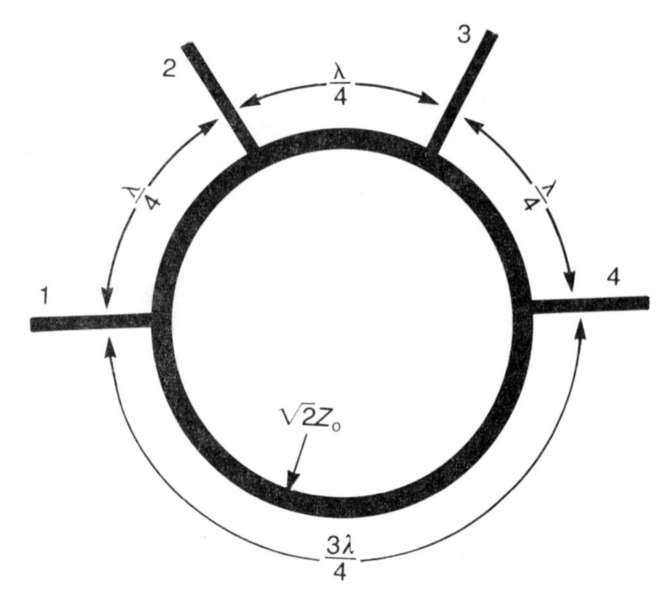

Fig. 74 – Rat-race.

impedance of $\sqrt{2}\ Z_o = 1.414\ Z_o$. The impedance of the four arms is $Z_o = 50\ \Omega$, and the relative spacing of the arms is shown in Fig. 74.

If power is applied to arm 1, then half of that power will appear at arm 2 and the other half at arm 4. The phase difference between arm 2 and arm 4 is 180°. If power is applied to arm 3 then one half of it appears in arms 2 and 4. The phase difference between the two arms is then 0°. For the two examples quoted there is isolation between arms 1 and 3, and for the second case between arms 2 and 4.

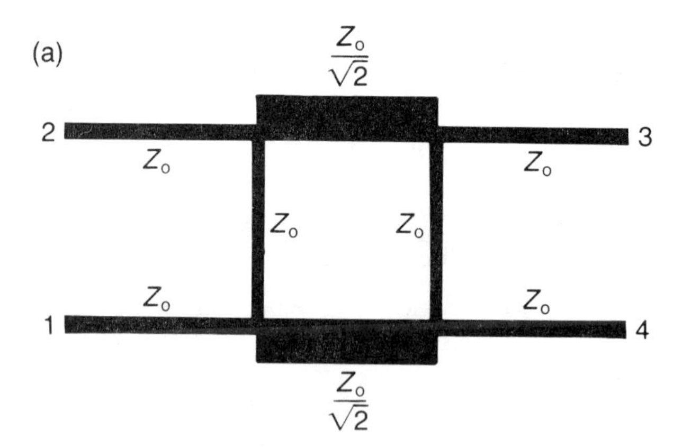

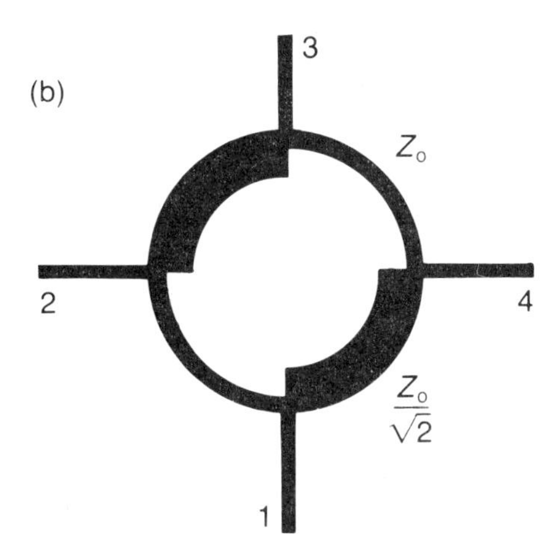

Fig. 75 – Branch line couplers.

Before implementing this design in a mixer, consider the operation of a hybrid coupler also known as 'branch line coupler'. There are many versions and configurations of this coupler. Two possible configurations are shown in Fig. 75. Power entering arm 1 will divide equally between arms 3 and 4 with a

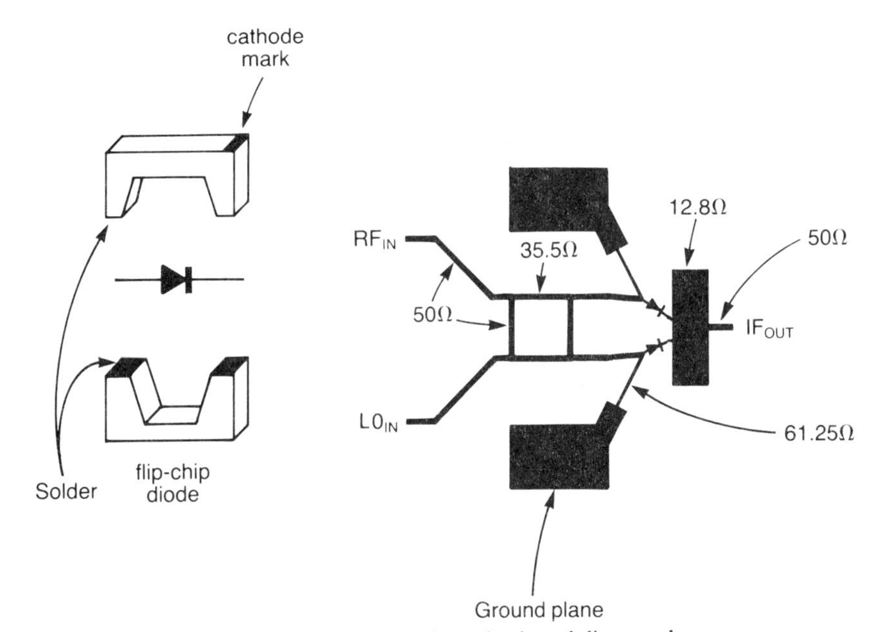

Fig. 76 – Mixer configuration using branch line coupler.

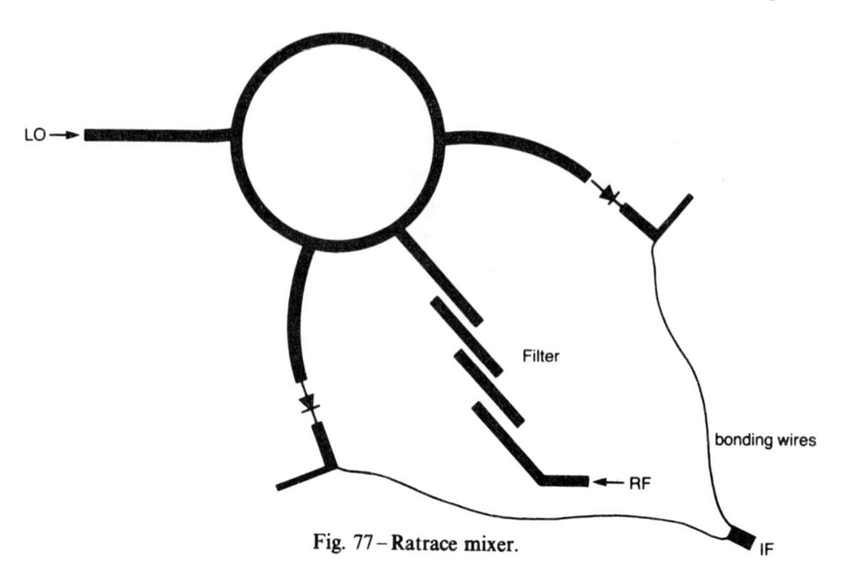

Fig. 77 – Ratrace mixer.

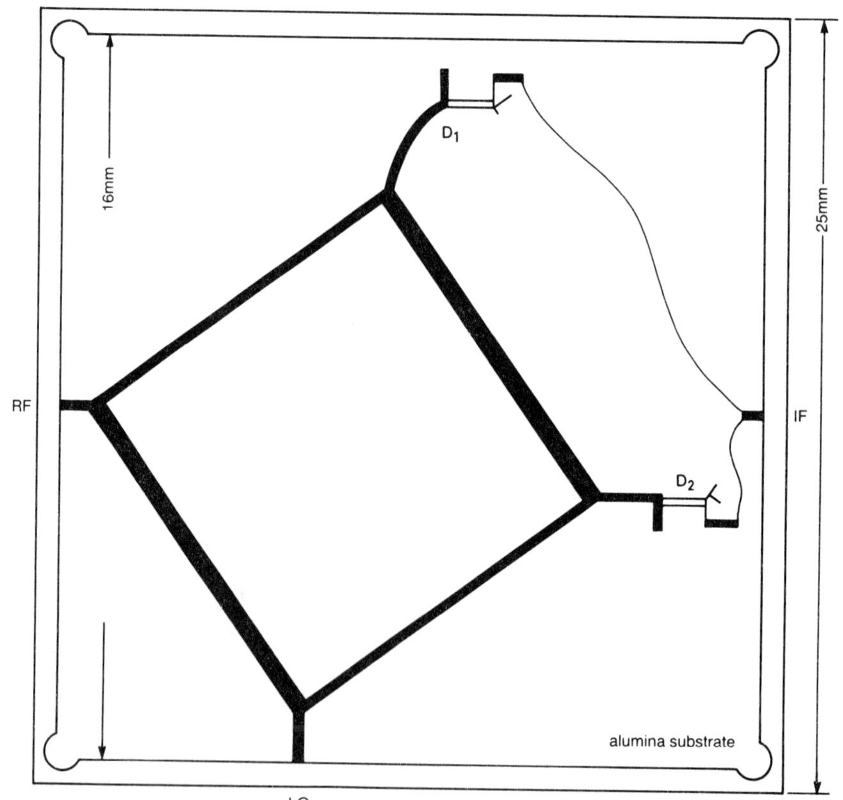

Fig. 78 – High-performance X-band mixer layout.

phase difference of 90° between these two arms. Arm 2 is isolated with respect to arm 1. The impedance of all arms is Z_0.

A typical practical mixer configuration using a branch line coupler is shown in Fig. 76. If the IF is high enough, the mixer diodes are connected at the common end to a filter laid out in microstrip. At higher microwave frequencies flip-flop diodes are used. A rat-race mixer configuration with a microstrip bandpass filter in the signal path is shown in Fig. 77. As a final example Fig. 78 is the schematic of a high performance X-band mixer with various refinements in terms of filtering.

6.3 LOCAL OSCILLATOR

The local oscillator forms an important part of frequency converter. When used in satellite communication and broadcasting systems the oscillator should be of simple structure and mechanically robust. Furthermore, the ideal oscillator would have most of the following features:

a) wide range of oscillation frequencies,
b) easy tuneability,
c) good frequency stability,
d) good frequency stability without hysteresis,
e) low noise properties,
f) high efficiency,
g) good output power.

To build an oscillator, passive or active devices may be used. Frequently a Gunn diode or a GaAsFET are employed. The following paragraphs give a brief explanation of a GaAsFET oscillator which is stabilised with a dielectric resonator, also known as DRO (dielectric resonator oscillator) [13, 14, 15, 16].

Ishihara *et al.* [15] have described an oscillator with the dielectric resonator in the feedback circuit. The microwave integrated circuit (MIC) schematic pattern of a DRO is shown in Fig. 79. Such oscillators are typically

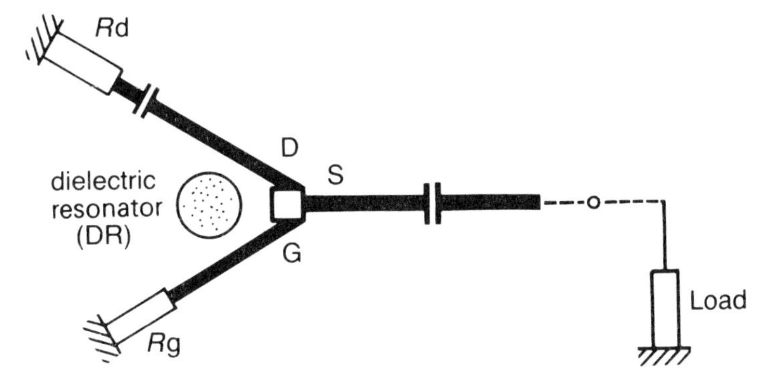

Fig. 79 – Schematic of a DRO.

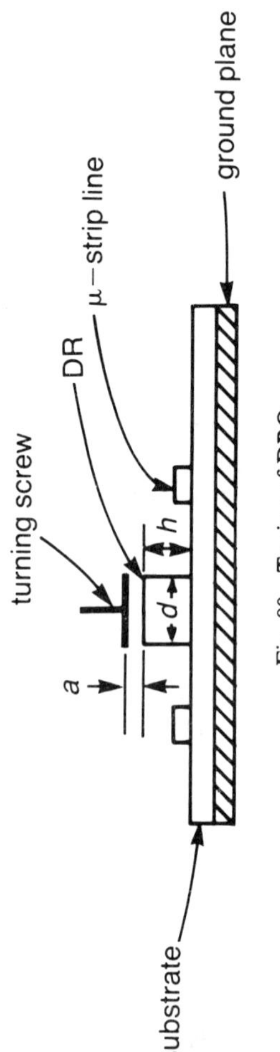

Fig. 80 – Tuning of DRO.

built on microwave substrates such as Alumina, Duroid, or Rexolite. The microstrip pattern is etched using standard photolithographic techniques. The DR is glued to the substrate as shown in Fig. 80. The operation of the circuit is as follows: microwave power incident on the gate is amplified and then fed back from the drain to the gate via the DR. The width of the drain and gate microstrip is such that a good match is provided between the FET and the respective resistor. Furthermore, the value of the resistors R_d, R_g and the angle between the drain and gate microstrip have a bearing on the microwave power generated. This power is taken from the source. A power in the order of 10 mW is usually required to operate a mixer.

The frequency of the DRO is mainly governed by the physical dimensions of the DR, i.e. its diameter d and its height h. The frequency can be changed by means of a tuning screw above the DR. For an 11 GHz oscillator frequency the DR is typically of 5.5 mm diameter and 2.1 mm height, and it has a dielectric constant ε_r of about 37. Provided that all other parameters are kept constant, the frequency increases as the height of the DR is reduced, and vice versa. These effects are shown in principle in Figs 81 and 82. Table 6.3 lists some commercially available dielectric materials. The data provided are approximate.

The DRO GaAsFET oscillator needs a dual-polarity bias supply. The positive voltage between 3 V and 7 V is required for the drain, and the negative voltage for the gate. It is, however, preferable to operate the oscillator from a single power supply. A typical bias circuits for this case is shown in Fig. 83. As can be seen, the gate is grounded, the source-to-gate voltage being provided by a resistor of about 12 Ω. To minimise source-to-gate voltage variations due to temperature, a resistor is connected between drain and source. The supply voltage is filtered by means of a feedthrough capacitor. A zener diode protects

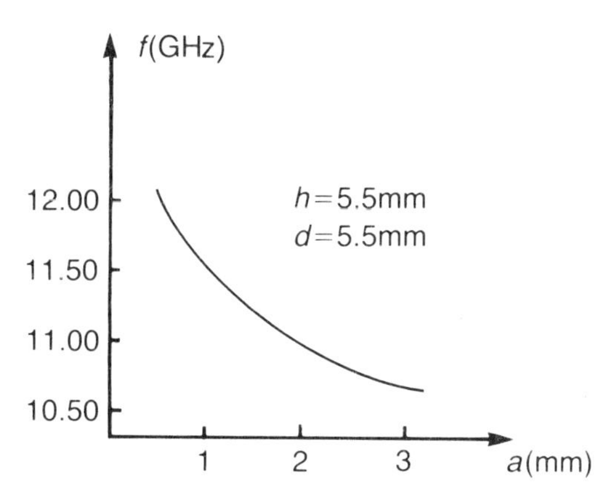

Fig. 81 – Mechanical tuning characteristic.

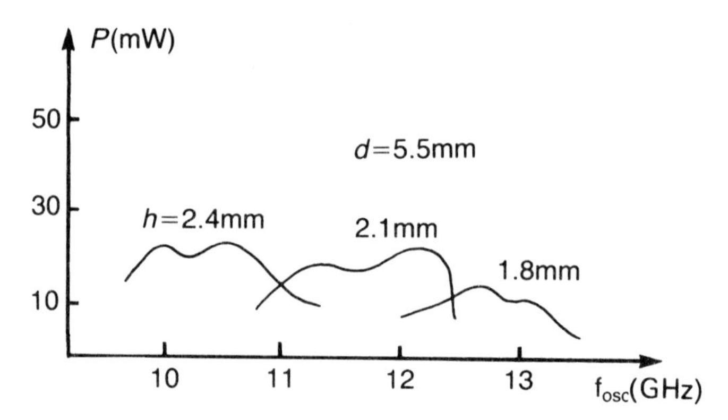

Fig. 82 – Mechanical tuning with different DRs.

Table 6.3 DR materials

Manufacturer	Material	ϵ_r	Q at 10 GHz
Ampex	Barium tetratitanate	38	3000
Murata/Erie	Zirconium tin titanate	38	4000
Siemens	Zirconium tin titanate	38	5000
Thompson CSF	Zirconium tin titanate	37	4000
Trans-Tech Inc	Zirconium tin titanate	37	4000
Marconi (MEDL)	Barium tetratitanate	38	4000

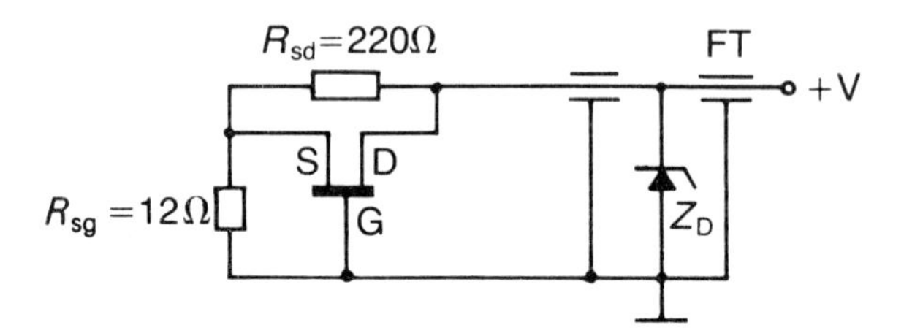

Fig. 83 – FET bias circuit for single power supply.

Table 6.4 Performance of microwave GaAsFET DRO for satellite TV

Oscillation frequency (in range of)	8–14 GHz
Output power	up to 70 mW
Mechanical tuning range	upto 1.5 GHz
Bias voltage	3–7 V
Efficiency	20 %
Frequency stability (– 20°C to 60°C)	± 200 kHz

the transistor from destruction by too high a supply voltage. Another example of a FET DRO is shown in Fig. 84, and a layout in Fig. 85.

The performance of a microwave GaAsFET DRO for satellite TV can be generally summarised as shown in Table 6.4.

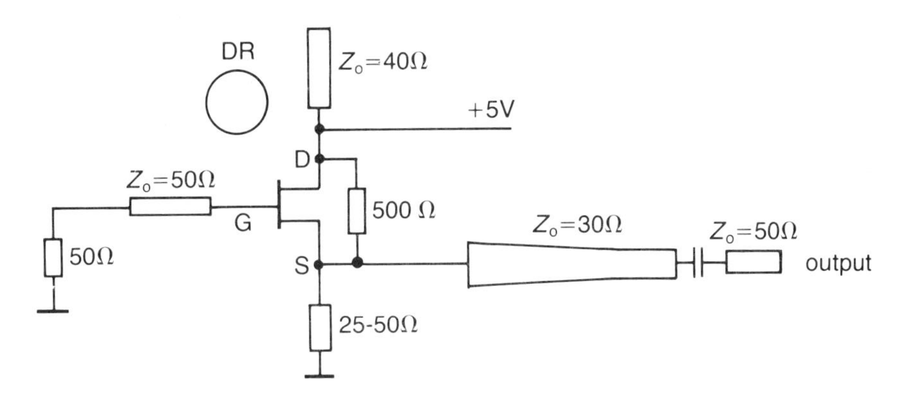

Fig. 84 – Layout and biasing of GaAs FET DRO.

6.4 IF AMPLIFIER

Once the signal has been converted down to a suitable frequency it needs further amplification by means of the IF amplifier which at the same time rejects the sum frequency produced by the mixer. The bandwidth of the IF stages must be wide enough to pass the modulated signal, and this governs to some extent the choice of the IF centre frequency. For satellite television a suitable first IF is 800 to 1500 GHz. The layout and dimensions for a 1.2 GHz IF amplifier are shown in Fig. 86. The only differences between this amplifier and the previously described microwave amplifier are the longer line lengths.

6.5 FM DEMODULATION FOR DBS RECEIVERS

Almost all existing television satellites use frequency modulation where bandwidth is exchanged for transmitted power. The reason is that conventional AM requires considerably higher transmitted power.

Frequency modulation is one of the possibilities which trades bandwidth for signal-to-noise ratio (SNR). FM signals are less susceptible to noise as long as the SNR at the input of the receiver is above a certain threshold which is dependent on the type of demodulator employed. Below the threshold previously referred to, the SNR of the demodulator output deteriorates rapidly. It is thus apparent that any extension of the threshold would result in a considerable advantage for the receiving system. For example, an improve-

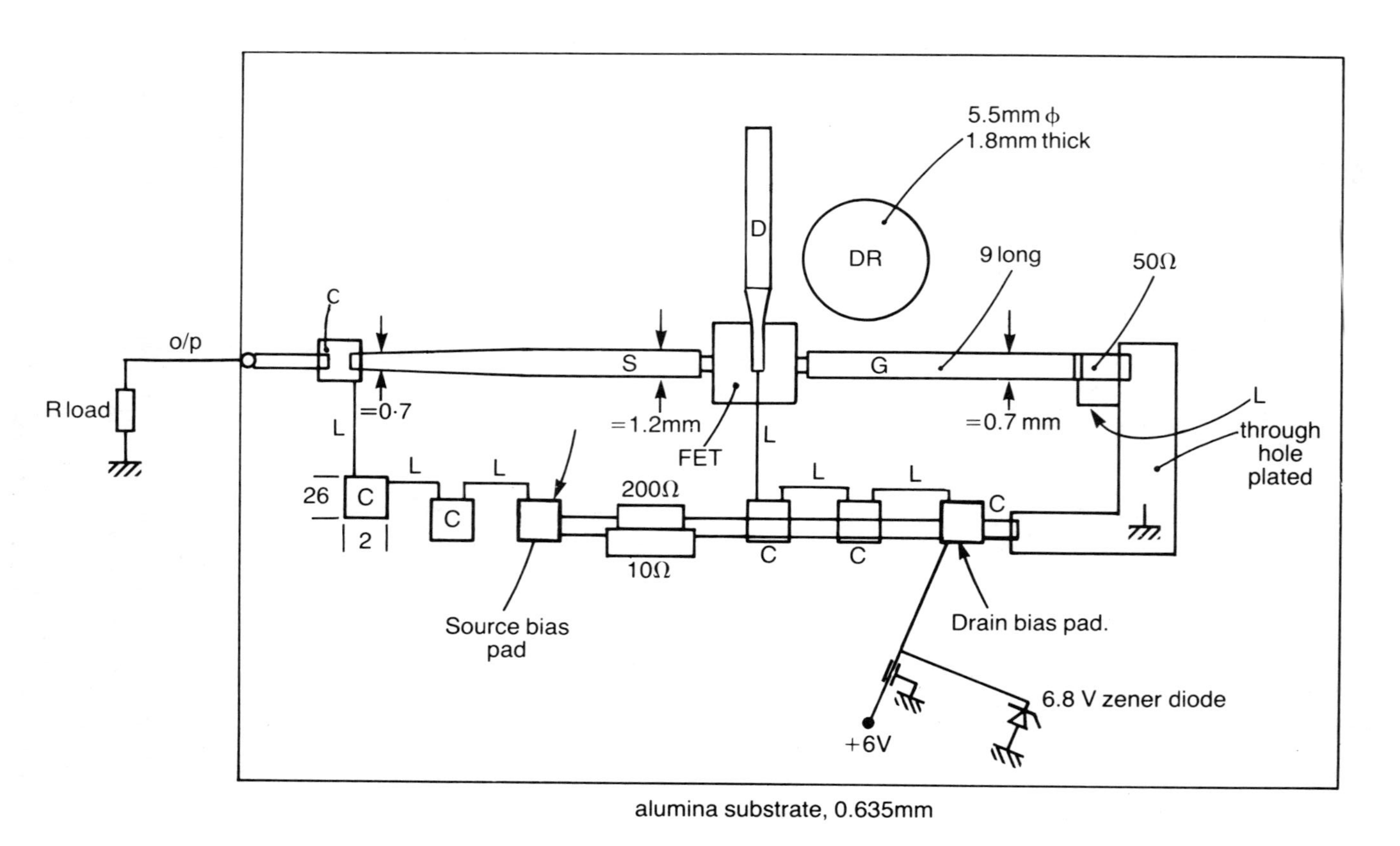

Fig. 85 – FET DRO; a practical layout on alumina substrate.

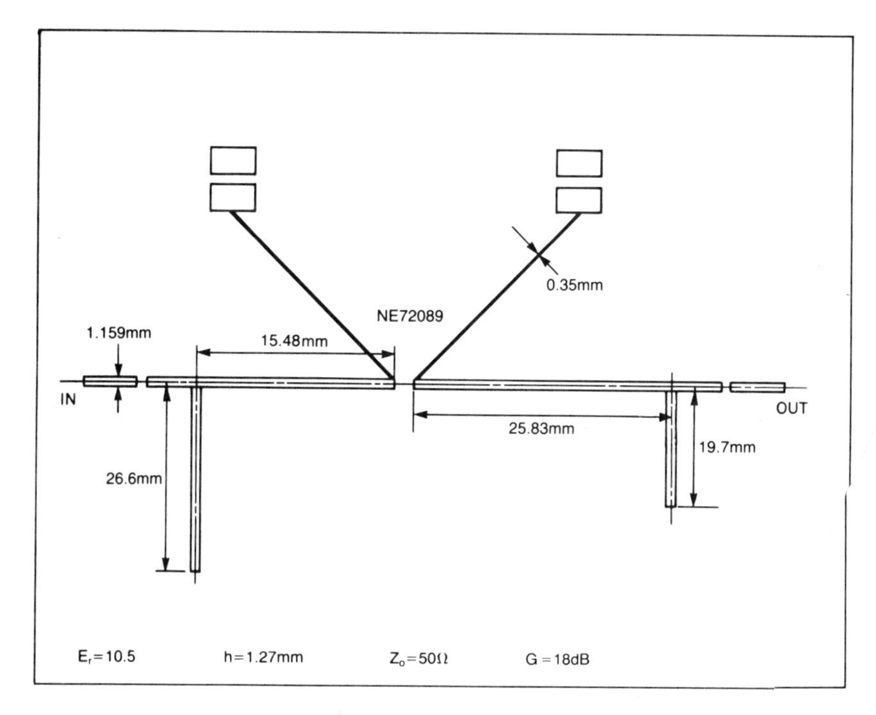

Fig. 86 – IF amplifier for 1.2 GHz.

ment in threshold by say 3 to 5 dB could allow for a cheaper antenna preamplifier (e.g. noisier or with less gain), or it could require a smaller antenna dish of 2 m rather than 3 m. From a consumer point of view, price and the mechanical simplicity of a direct broadcast satellite receiver are of great importance. The following sections describe methods which allow threshold extension [17]. The techniques for lowering the threshold in FM receivers fall in three categories:

a) the dynamic tracking filter (DTF),
b) the FM demodulator with feedback (FMFB, FCFB),
c) the phase locked loop demodulator (PLL).

The fundamental difference between a conventional FM demodulator and an extended threshold demodulator is that the latter uses an important piece of 'a priori' information: even though the carrier frequency can have large deviations, the rate of change of those frequency deviations takes place at the baseband frequency. All threshold extension schemes are basically tracking filters which track only the slowly varying frequency of the modulated carrier. Thus they respond only to a narrow band around the instantaneous carrier frequency.

6.5.1 The dynamic tracking filter

The bandwidth which is required by a demodulator can be reduced by either reducing the signal deviation or by tracking the signal with a reduced bandwidth filter. A DTF scheme is shown in Fig. 87. Threshold extension is achieved by reducing the noise power available to the demodulator input. Once designed, the filter bandwidth remains unchanged. Filter bandwidth reduction must not fall below a certain limit as this will cause mistracking and an eventual loss of signal. The idea of the DTF comes from the USSR (about 1950). The TELSTAR receiver at Goonhilly in Cornwall employed a DTF. Varactors are used in the filter to facilitate tracking. As shown in Fig. 87, a control voltage is derived from the demodulated baseband frequency of the filter. The resonance frequency is thus made to follow the instantaneous variations of the carrier. For a hyper-abrupt diode the capacitance is inversely proportional to the square of the voltage, i.e. $C \propto V^{-2}$. This applies for a large part of its characteristic. Since the filter resonance frequency is proportional to $1/LC$, the filter can track linearly with applied voltage over a large range. Although the DTF employs a feedback loops it is not a negative feedback loop. Hence, the loop must have an extremely high order of linearity and must have a reasonably large bandwidth. There are two more factors to be considered in the filter design. When tracking, the filter should have a minimum of amplitude variation over the tracking range, and it should have a constant bandwidth when tracking.

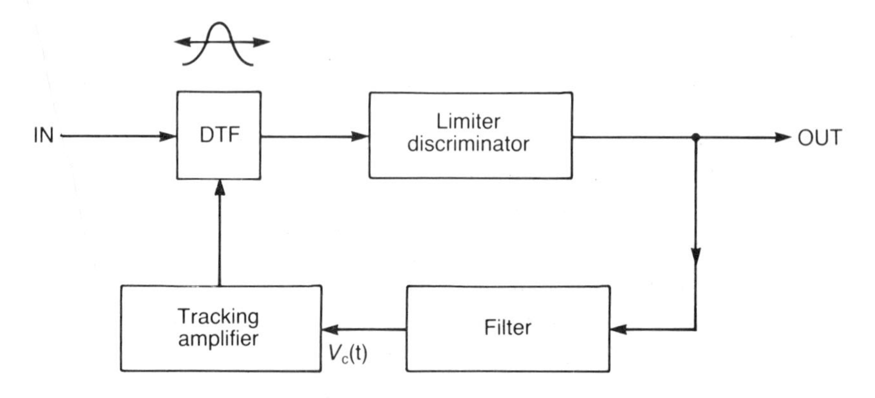

Fig. 87 – Blockdiagram of dynamic tracking filter (DTF).

6.5.2 Frequency modulated feedback

The idea of FMFB stems from Chaffee (1939). With the advent of space exploration interest in FMFB was revived. One of the first uses was in project ECHO and was described by Ruthroff. The block diagram of an FMFB demodulator is shown in Fig. 88. To reduce the threshold, the bandwidth of

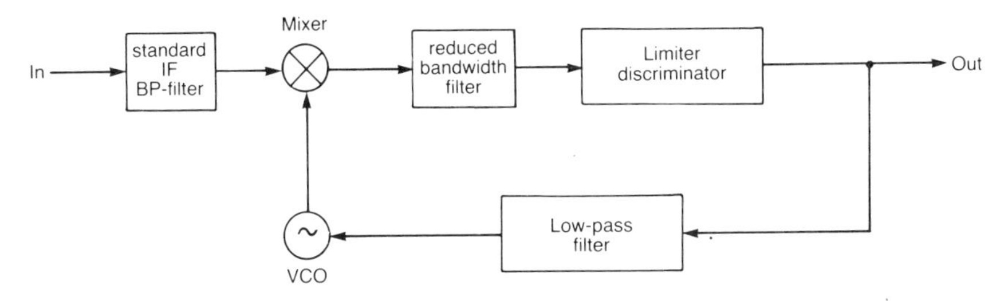

Fig. 88 – Block diagram of FMFB circuit.

the demodulator is reduced by reducing the input signal deviation. This is done by taking the demodulator output to modulate a voltage controlled oscillator (VCO) in such a manner that the VCO frequency follows, but does not coincide with, the instantaneous frequency of the incoming IF. This results in a compression (hence the term FCFB) of the incoming wideband IF into a narrower band prior to demodulation. It is important to note that threshold extension is not caused through the use of feedback, but solely by the reduction of bandwidth made possible by the feedback. For feedback without bandwidth reduction no threshold extension will result. Vice versa, bandwidth reduction without feedback will result in distortion.

6.5.3 The phase-locked loop

The phase locked loop [18, 19, 20, 21] has become a major circuit for use in communication engineering. Its uses embrace the fields of signal acquisition, frequency synthesis, synchronisation, and last but not least, FM demodulation. The diagram of a phase locked loop (PLL) is shown in Fig. 89. The basic PLL consists of a phase detector, low-pass filter, and a voltage controlled oscillator (VCO). If the phase between oscillator and RF input

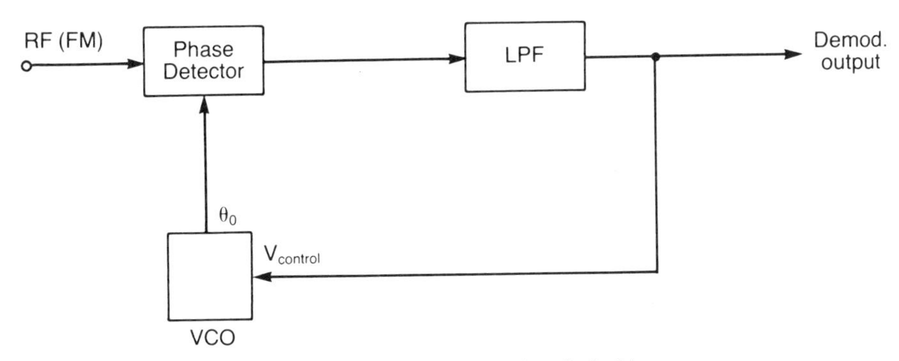

Fig. 89 – Block diagram of a phase-locked loop.

changes, indicating that the input frequency is changing, then the phase detector output changes. The polarity of the phase detector output depends on its design and whether the FM signal is compressing or expanding. The detector output is then low-pass filtered and the average fed to the VCO, which will adjust its frequency to near that of the incoming RF signal. Lock is thus preserved. From this it is evident that the low-pass filter output which acts as a control voltage for the VCO is also the demodulated RF.

The threshold of a PLL is heralded by clicks (about one click per second) at its output, similar to conventional FM demodulators. Considerable care has to go into the design of the low-pass filter, especially as it determines the stability of the loop. The PLL offers a performance which is comparable to that of the DTF and FMFB. One of its advantages is its simplicity.

6.6 PHASE-LOCKED LOOP COMPONENTS

Demodulators as used for FM reception are characterised by a characteristic as shown in Fig. 90. This curve displays the quality of the signal at the output of the demodulator as a function of the signal quality/purity at the input. The curve consists of three distinct regions:

a) the distortion limited region,
b) the FM improvement region,
c) the region below threshold.

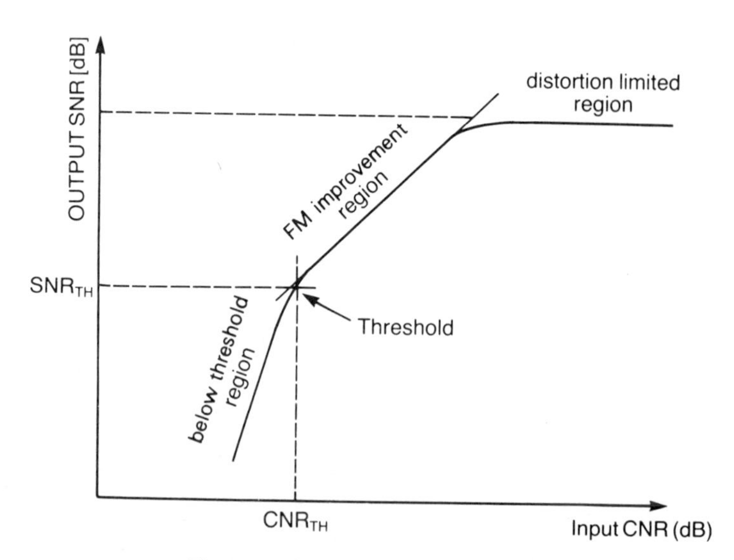

Fig. 90 – FM demodulator characteristics.

For high carrier-to-noise ratios (CNR) the output signal fidelity is limited by the distortions of the systems, hence the distortion limited region. Under normal conditions the system works in the FM improvement region. The noise originates from the fact that reception quality can be improved at the expense of bandwidth (for a given transmitter power). Below threshold, i.e. in the below threshold region, the signal quality at the output of the demodulator deteriorates rapidly.

6.6.1 The demodulator

The basic and most important element in an FM receiver is the demodulator. Any improvements obtained with other FM schemes are measured against this basic element. A block diagram of a typical FM receiver is shown in Fig. 91. The composition of a conventional demodulator can be explained in the following manner: the demodulator is sandwiched between a pre-detection filter and a post-detection filter in order to keep noise to a minimum. The demodulator consists of (theoretically at least) three parts, namely a limiter, a differentiator, and an envelope detector. The limiter removes amplitude variations which may otherwise be converted into FM variations. The differentiator 'detects' the frequency variations, i.e. it transforms the FM into AM, whilst the baseband is finally recovered by an envelope detector.

In a practical circuit it may not be possible to subdivide the demodulator into three different elements. Nevertheless, it may help in the appreciation of its operation and to locate possible reasons for maloperation.

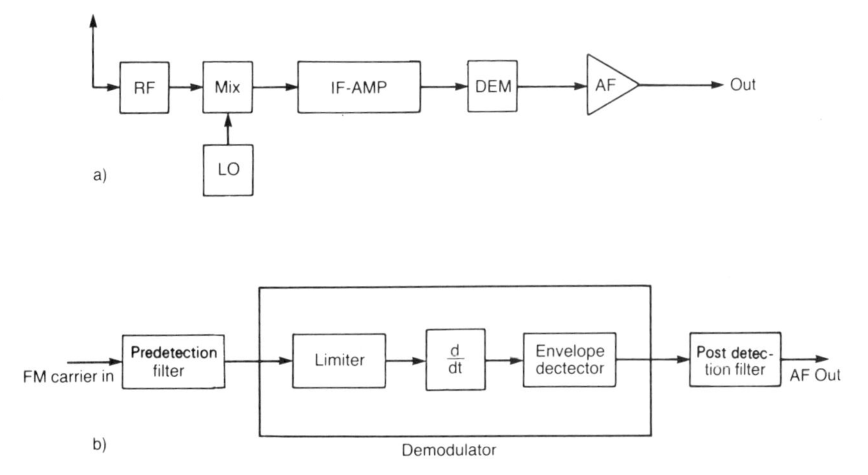

Fig. 91 – Typical FM receiver (a) and conventional FM demodulator (b).

6.6.2 Filtering of FM carrier

In FM threshold extended schemes two filters are employed, a broadband filter (BB) and a narrow-band filter (NB). The BB filter filters the wave before it is applied to the mixer (phase detector) or to the loop (in case of FMFB). The NB filter is incorporated within the loop of an FMFB system. It is useful to know the distortion created by the filter since this will give an indication when the distortion limited region commences. Phase linearity of the filter is also important, since non-linearity causes distortion products. The filter bandwidth is largely governed by the modulation indices encountered. For analogue signal transmissions Carson's rule is a useful rule of thumb. The filter within a FMFB loop is usually a single-tuned circuit.

6.6.3 The local oscillator and VCO

Any FM receiver requires an oscillator for signal translation. For threshold extension circuits a voltage controlled oscillator (VCO) is required. This is an oscillator whose instantaneous frequency is changed in accordance with an applied control voltage. Ideally, the oscillator frequency will vary linearly around its centre frequency with applied voltage. This is not easily achieved, especially for large deviations of ± 13.5 MHz as in the case of satellite TV.

The VCO has to satisfy requirements which may vary from application to application. Furthermore, some of the requirements are interrelated, and some conflict with each other. The requirements are:

a) linear frequency deviation with control voltage,
b) stability of frequency and phase,
c) spectral purity,
d) modulation bandwidth,
e) sensitivity.

6.6.4 Realisation of PLL demodulator

A more detailed block diagram of a PLL threshold extension demodulator is given in Fig. 92, and a circuit diagram in Fig. 93. A mixer is being used as a phase detector.

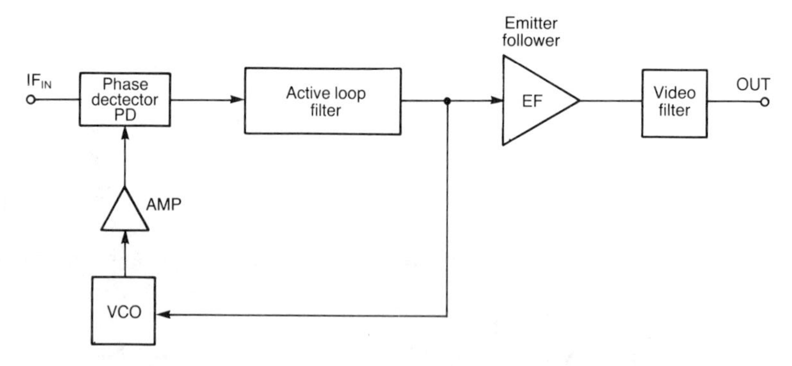

Fig. 92 – Block diagram of threshold extension demodulator using PLL.

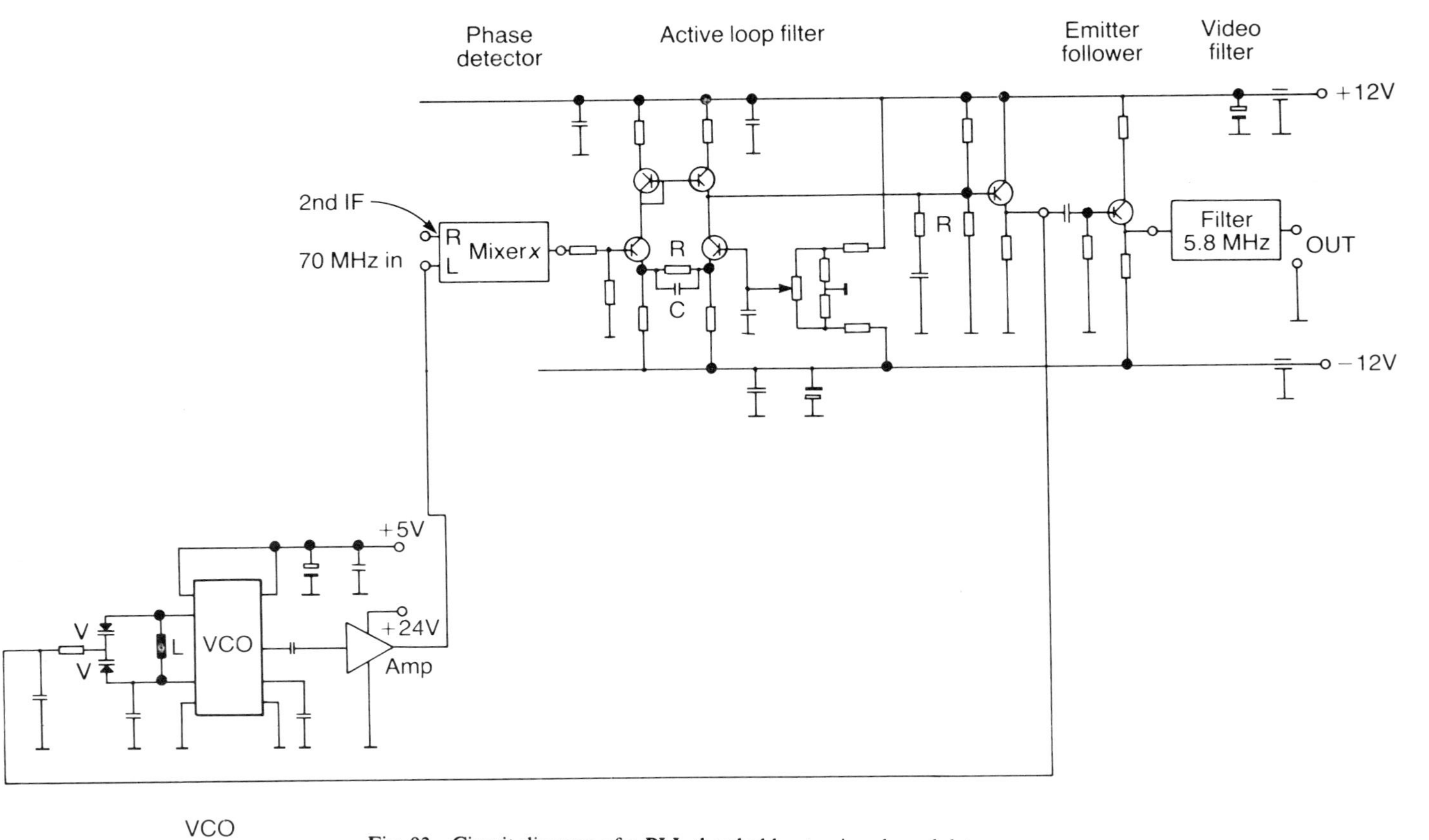

Fig. 93 – Circuit diagram of a **PLL** threshold extension demodulator.

The VCO employs a high-speed flip-flop. The two varactors V together with the inductor L form a frequency selective network. If the second IF of a DBS receiver is 70 MHz, then these components must be chosen together with a bias voltage to resonate at 70 MHz. In Fig. 93 the bias or control voltage is taken from the output of the loop filter. If the loop filter cannot provide the right bias voltage then some kind of level shifting must be introduced. To provide minimum power conversion loss the mixer (phase detector) must be driven by about + 7 dBm. This can be done by boosting the VCO output with a suitable wideband amplifier. The phase detector output X is passed through an active loop filter consisting of a differential amplifier. The gain of this stage is governed by the size of the coupling emitter resistance. This resistor, together with the parallel capacitor, governs the first break point of the filter characteristic. The base of the emitter follower within the loop filter is connected via an RC combination to ground. This determines the second break point. The demodulated signal is taken via a second emitter follower and a filter to the TV monitor.

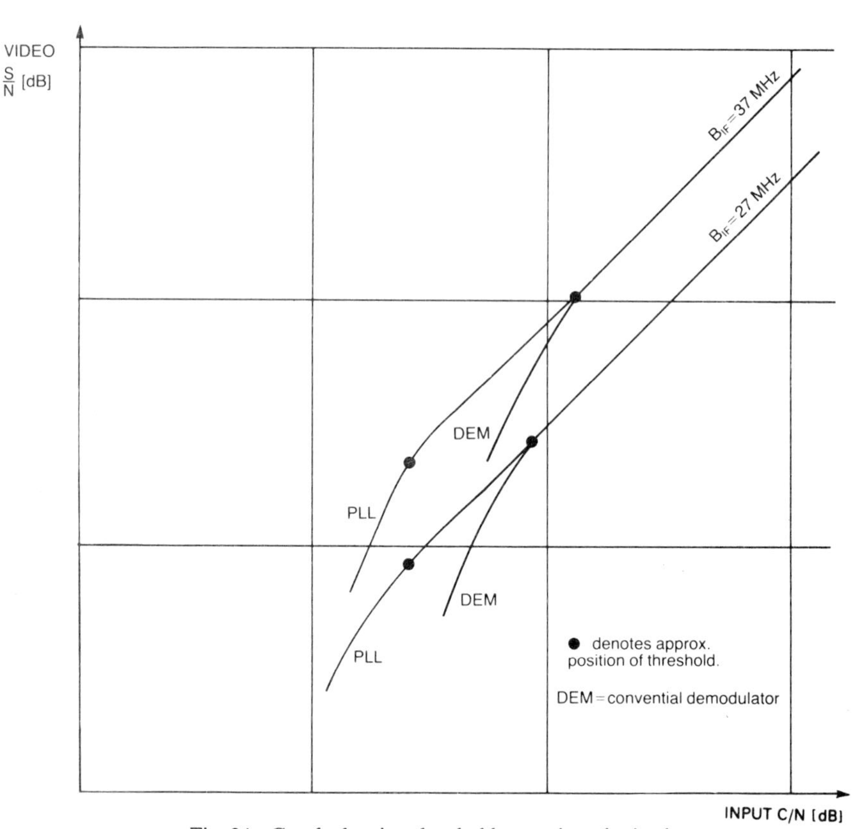

Fig. 94 – Graph showing threshold extension obtained.

A graph showing threshold extension, which is typically about 3 dB, is shown in Fig. 94. The coordinates have not been numbered, since the measuring technique employed should first be specified. The threshold for each type of demodulator is given approximately. Again, the criteria for threshold definition should be given.

REFERENCES

[1] Low-noise GaAsFET amplifiers. *Watkins-Johnson Company, Technotes*, **4,** No. 1, Jan./Feb. 1977.

[2] A. Russel, Amplifier application notes. *Microwave Journal*, **25,** No. 10, Oct. 1982.

[3] D. W. Maki *et al.*, Monolithic low-noise amplifiers. *Microwave Journal*, **24,** No. 10, Oct. 1981.

[4] Tri T. Ha, *Solid state microwave amplifier design*. John Wiley and sons, New York, 1981.

[5] S. R. Tindale & L. A. Trinogga, The analysis and design of single stage microwave semiconductor amplifiers using microstrip transmission lines. *Project report*, April 1983.

[6] *S*-parameters – circuit analysis and design. *Hewlett Packard application note* 95,1968.

[7] T. H. Edgar, Investigation of low-noise microwave receivers based on the theoretical analysis of local oscillator noise contribution. PhD thesis, Newcastle upon Tyne Polytechnic, 1980.

[8] S. W. Holland, Modelling and measurement of mixer diodes working at mm wave frequencies. M.Phil. thesis, Newcastle upon Tyne Polytechnic, 1982.

[9] D. J. Gunton, Microwave balanced mixer circuits using combline directional couplers. *Electronics letters*, **12,** No. 8, April 1976, pp. 204.

[10] B. Henderson, Mixer design considerations improve performance. *Microwave system news*, Oct. 1981.

[11] Chen Y. Ho *et al.*, Half-wave length and step impedance resonators aid microstrip filter design. *Microwave system news*, **13,** No. 10, Oct. 1983, pp.71.

[12] R. S. Sleven, Pseudo exact band pass filter design saves time. Part 4 *Microwaves*, Dec. 1968.

[13] C. Tsironis *et al.*, Temperature stabilisation of GaAs MESFET oscillators using dielectric resonators. *IEEE, MTT31*, No. 3, March 1983, pp. 312.

[14] Hiroyuki *et al.*, A highly stabilised low-noise GaAs FET integrated oscillator with a dielectric resonator in the C-band. *IEEE, MTT26*, No. 3, March 1978, pp. 156.

[15] Osamu Ishihara *et al.*, A highly stabilised GaAs FET oscillator using a

dielectric resonator feedback circuit in the 9–14 GHz. *IEEE, MTT28*, No. 8, Aug. 1980, pp. 817.

[16] J. K. Plourde, Application of dielectric resonatoes in microwave components. *IEEE, MTT29*, No. 8, Aug. 1981, pp. 754.

[17] L. A. Trinogga, TV-FM demodulator threshold extension. *IBA research report*, Oct. 1981.

[18] R. Citta, Frequency and phase locked loop. *IEEE Transactions*, **CE-23,** No. 3, Aug. 1977, pp. 358.

[19] B. N. Biswas, Phase locked demodulator threshold: a new criterion *IEEE Transactions*, **AES 12,** No. 5, Sept. 1976, pp. 537.

[20] A. Blanchard, Phase locked loop behaviour near threshold. *IEEE Transactions*, **AES 12,** No. 5, Sept. 1976, pp. 628.

[21] F. M. Gardner, *Phase locked loop techniques*. Wiley and sons, New York, 1966.

7

Satellite receiver designs

7.1 PLANAR TRANSMISSION LINES

Some of the frequencies relating to DBS reception are very high, i.e. fall into the microwave region. Clearly, special techniques have to be employed to process those frequencies. Planar transmission lines [1, 2] are most suitable for microwave frequencies since the lines have a geometry which allows a definition of the line impedance by controlling its dimensions on one plane. This feature allows not only single lines to be produced but complete microwave circuits and subsystems. Fabrication is by photolithographic techniques, similar to that of printed circuit boards.

Various configuration of planar transmission lines are possible, e.g. microstrip, stripline, and slotline. The microstrip is most commonly used for microwave circuits as employed for DBS receivers, and is thus treated in more detail.

A cross-sectional view of a microstrip is shown in Fig. 95. The dielectric substrate has a ground plane on one side and the strip on the other side. This results in one of the advantages of microstrip in relation to other transmission

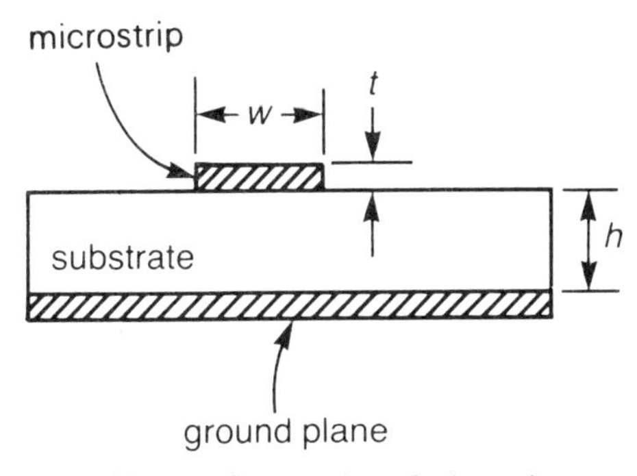

Fig. 95 – Cross-section of microstrip.

lines, namely easy access to the strip. This allows easy mounting of active and passive devices. Furthermore, microstrip offers itself for mass production. At lower microwave frequencies the substrate is usually covered with copper only.

Commercially available materials are known by their trade names of Rexolite, RT-Duroid, and Polyguide. For higher microwave frequencies, better quality substrates are used such as Alumina. To provide high quality strips, the copper sides are coated with chromium and gold. Examples of microwave substrates and some of their properties are given in Table 7.1.

Table 7.1 Examples of commonly used microwave substrates

Name	ε_r	thickness (h)
Rexolite	2.4	varies from
Polyguide	2.36	0.635 mm
RT Duroid	2.2–10.5	to
Alumina	9.6–10.2	1.5875 mm

The impedance and length of a microstrip are predominantly governed by the thickness h and permittivity ε_r of the substrate and the width w of the microstrip. Detailed expositions on the subject can be found in the appropriate literature. The computer programme 'STRIP' which is listed in the Appendix uses the following approximate synthesis equation for the calculation of microstrip width w:

$$\frac{w}{h} = \left(\frac{e^H}{8} - \frac{1}{4e^H}\right)^{-1} \tag{7.1}$$

where

$$H = \frac{Z_o\sqrt{2(\varepsilon_r + 1)}}{119.9} + \frac{\varepsilon_r - 1}{2(\varepsilon_r + 1)}\left(\ln\frac{\pi}{2} + \frac{1}{\varepsilon_r}\ln\frac{4}{\pi}\right). \tag{7.2}$$

The program can be used as follows. Obtain a data sheet for the substrate in question, say RT Duroid 6010. The material is available with a permittivity of 10.5 and thickness of 0.025 in. = 0.635 mm and 0.05 in. = 1.27 mm. If we want to produce a microstrip of 50 Ω impedance (or 100 Ω) and input the appropriate data, then we obtain the computer printout as on page 143.

From this we see that the line width decreases as the impedance increases. For a substrate thickness of 0.635 mm we obtain the result as shown on p. 143.

In relation to the previous results this shows that line width decreases as the substrate thickness decreases. Hence the line width can be manipulated by choosing appropriate substrate thickness and permittivity. Table 7.2 lists some line widths and impedances which are useful for the design of the microwave circuitry discussed in earlier sections.

```
INPUT DATA
Z0=50 Ohm
ER=10.5
H=1.27 mm

RESULT
W/H=0.91261898
since H=1.27 mm
thus W=1.1590261 mm
EEFF = 7.0127902
```

```
INPUT DATA
Z0=50 Ohm
ER=10.5
H=0.635 mm

RESULT
W/H=0.91261898
since H=0.635 mm
thus W=0.57951305 mm
EEFF = 7.0127902
```

```
INPUT DATA
Z0=100 Ohm
ER=10.5
H=1.27 mm

RESULT
W/H=0.12051522
since H=1.27 mm
thus W=0.15305434 mm
EEFF = 6.2236462
```

```
INPUT DATA
Z0=100 Ohm
ER=10.5
H=0.635 mm

RESULT
W/H=0.12051522
since H=0.633 mm
thus W=.076527188 mm
EEFF = 6.2236462
```

Table 7.2 Useful line widths and impedances (RT Duroid 6010)

Z_o (Ω)	h (mm)	ε_{eff}	ε_r	w (mm)	h (mm)	ε_{eff}	ε_r	w (mm)
30	1.27	7.64	10.5	2.86	0.635	7.64	10.5	1.43
35.35	1.27	7.43	10.5	2.20	0.635	7.43	10.5	1.10
50	1.27	7.01	10.5	1.15	0.635	7.01	10.5	0.57
70.7	1.27	6.59	10.5	0.49	0.635	6.59	10.5	0.28
100	1.27	6.22	10.5	0.15	0.635	6.22	10.5	0.07
150	1.27	5.92	10.5	0.02	0.635	5.92	10.5	0.01

As can be seen from the table, line impedances above 100 Ω have impracticable width. Choosing a substrate with lower permittivity, e.g. 2.2, results in a wider line as shown in the following print-out.

```
INPUT DATA
Z0=35.35 Ohm
ER=2.2
H=1.27 mm

RESULT
W/H=5.374896
since H=1.27 mm
thus W=6.8261179 mm
EEFF = 1.9337146
```

```
INPUT DATA
Z0=50 Ohm
ER=2.2
H=1.27 mm

RESULT
W/H=3.1203287
since H=1.27 mm
thus W=3.9628175 mm
EEFF = 1.8725655
```

Two more points have to be considered when designing with microstrip. Firstly, the permittivity is a function of the width-to-height ratio (w/h) of the substrate, i.e.

$$\varepsilon_{eff} = \frac{\varepsilon_r + 1}{2} + \frac{\varepsilon_r - 1}{2}\left(1 + \frac{10h}{w}\right)^{-1/2}, \tag{7.3}$$

where ε_r is the relative permittivity as given on the manufacturers' data sheet, and ε_{eff} is the effective permittivity as used in calculations. Secondly, waves travelling through a dielectric medium are shortened in relation to their travel through air according to the following equation:

$$\lambda_{diel} = \frac{\lambda_{air}}{\sqrt{\varepsilon_{eff}}} \tag{7.4}$$

or

$$\lambda_{diel} = \frac{c}{f\sqrt{\varepsilon_{eff}}} \tag{7.5}$$

where λ is the wavelength in either air or the dielectric, f is the frequency of operation, and $c = 3 \times 10^{10}$ cm/s, the speed of light. For example if $\varepsilon_{\text{eff}} = 1$ then this means that air is the dielectric and $\lambda_{\text{diel}} = \lambda_{\text{air}}$.

For a material with $\varepsilon_{\text{eff}} = 9$, the wavelength would be one third of that in air. For a frequency of 10 GHz and $\varepsilon_{\text{eff}} = 9$ the wavelength in the substrate is

$$\lambda_{\text{diel}} = \frac{3 \times 10^{10}\ \text{cm s}}{10 \times 10^{9} \sqrt{9}\ \text{s}} = 1\ \text{cm}.$$

As stated earlier, a more accurate and exhaustive treatment of this subject can be found in the literature. Nevertheless, the information provided gives quick results up to frequencies at X-band.

7.2 EXAMPLES OF RECEIVER DESIGNS

Microwave signals at 12 GHz are rapidly attenuated as they pass along cables or waveguides, so it is important in all designs of receiving installations to ensure that the 12 GHz signals are converted as quickly as possible to a lower frequency so that attenuation and noise will be minimised. This lower frequency is generally known as the 'first IF' (intermediate frequency) and is usually around 1.3 GHz.

In 1977 there was considerable doubt as to whether it would be possible to achieve any amplification at 12 GHz in low-cost domestic units. Most of the prototype receiver designs that were published did not have a 12 GHz RF amplifier, but fed the incoming signal straight into a diode mixer which combined it with the signal from a Gunn-diode local oscillator. Such prototype designs achieved noise figures of 6–8 dB. One suggested design [3] that came from the research laboratories of Japanese broadcasters NHK used a metal sheet with holes cut in it, carefully folded so that it formed a waveguide frequency converter that looked as though it would be ideal for mass production, although it was never built in any quantities.

Radio engineers have long recognised the signal-to-noise ratio advantages that accrue if a good radio-frequency amplifier is used as the first stage of a receiver, and the development of relatively low-cost 12 GHz amplifiers has recently become practicable because of the parallel developments that have taken place in the production of semiconductor devices, and in particular, the Gallium-arsenide field-effect transistor, commonly known as a GaAsFET [4, 5, 6, 7, 8].

A typical receiver system consists of four main parts, as shown in Fig. 96. The first essential is the parabolic dish that we have considered at some length, and this is connected directly to the 'outdoor unit' which contains the low-noise RF amplifier and down-converter. A high-quality low-loss coaxial cable then feeds the signals down to the living room, where it connects to the 'indoor unit', an adaptor box containing the tuning unit permitting individual

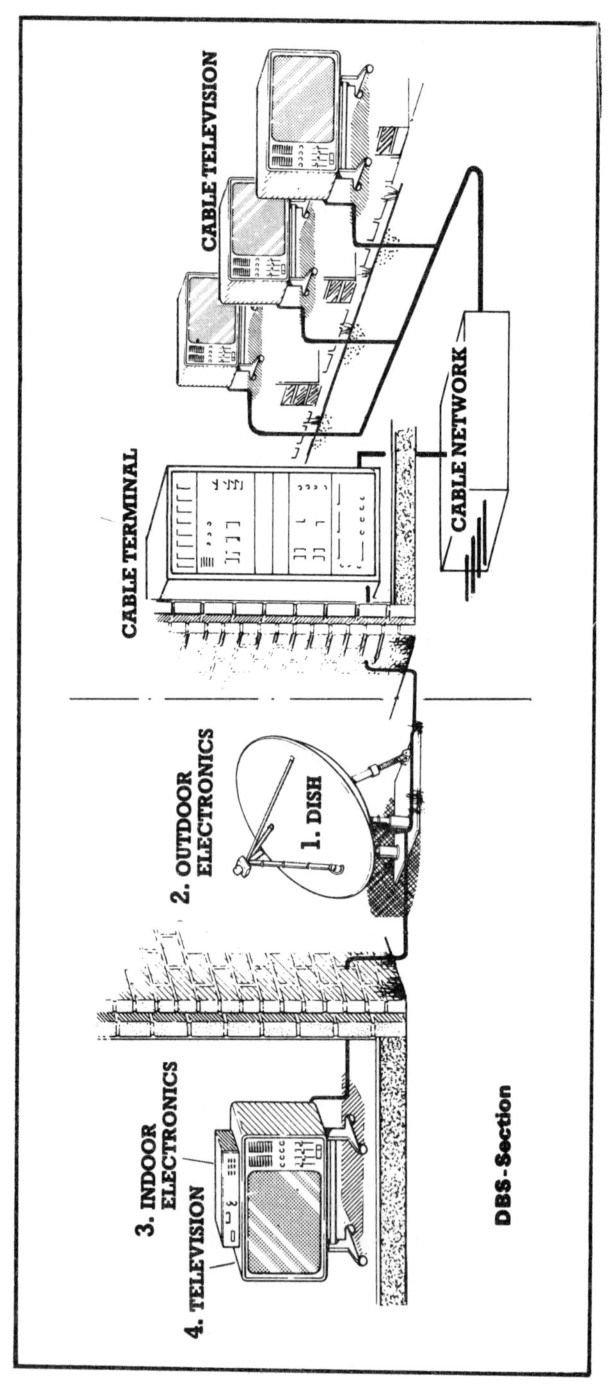

Fig. 96 – Typical DBS receiver configuration.

programmes to be selected, and usually a radio-frequency modulator which allows the signal to be fed into normal UHF television receivers through the aerial socket. For cable television systems the outdoor electronics connects to the cable terminal.

In systems using the Multiplexed Analogue Components (MAC) type of transmissions the indoor unit also contains the integrated circuits necessary to convert the incoming signals into the individual luminance and chrominance signals which can then be fed directly into the video stages of a modern receiver to provide the very highest quality pictures. As a somewhat temporary measure it can be arranged for the MAC signals to be coded into PAL and remodulated so that they can be fed into the aerial socket of older receivers, but this negates most of the advantages that the MAC system was designed to provide. The indoor unit will also contain the complex circuitry necessary to decode the multiple digital sound and data transmissions that are to be transmitted along with most European television transmissions.

The Japanese public broadcasting authority, NHK, began the first regular service of high-power direct broadcasts from satellites during 1984, using the BS-IIa (Fig. 97) satellite situated some 36 000 km above Borneo. Various dish aerials have been made available for this service. Viewers in the central areas of Japan are obtaining satisfactory reception with 60 cm diameter dishes, and most others are getting good results with either 75 cm or 1 m dishes.

A more detailed diagram of a good DBS receiver is shown in Fig. 98. The satellite signal is fed from the aerial into a low-noise microwave amplifier. Its purpose is to amplify the signal with as little noise addition as possible. Two stages are usually employed. The signal is then converted in a mixer into the first intermediate frequency which, in this case, is 750 MHz. The mixer is driven by an oscillator whose frequency must be very stable. As is well known, the sum and the difference frequencies of the signal and local oscillator are present at the mixer output. An intermediate frequency amplifier and filter are used to further amplify the signal and to define the bandwidth of the receiver.

The first IF is then applied to the input of the second mixer which, when driven with a local oscillator frequency of 680 MHz, will produce a second IF of 70 MHz. This frequency is generally accepted as a standard IF. Since two mixer stages are employed to convert the microwave signal to 70 MHz, the receiver is of the double heterodyne type. Direct conversion from the microwave frequency to 70 MHz could be done if it were possible to produce very highly stable oscillators. Various types of oscillators can be used to drive the second mixer. A crystal oscillator can be used as a primary source and its frequency be doubled and/or trebled to give 680 MHz. Alternatively, a solid state cavity controlled PLL referenced oscillator can be employed giving high frequency stability. The second IF is preamplified and fed into a filter which determines the IF bandwidth. This bandwidth is in the order of 27 MHz. The preamplifier must be linear and have a large dynamic range. The purpose of the

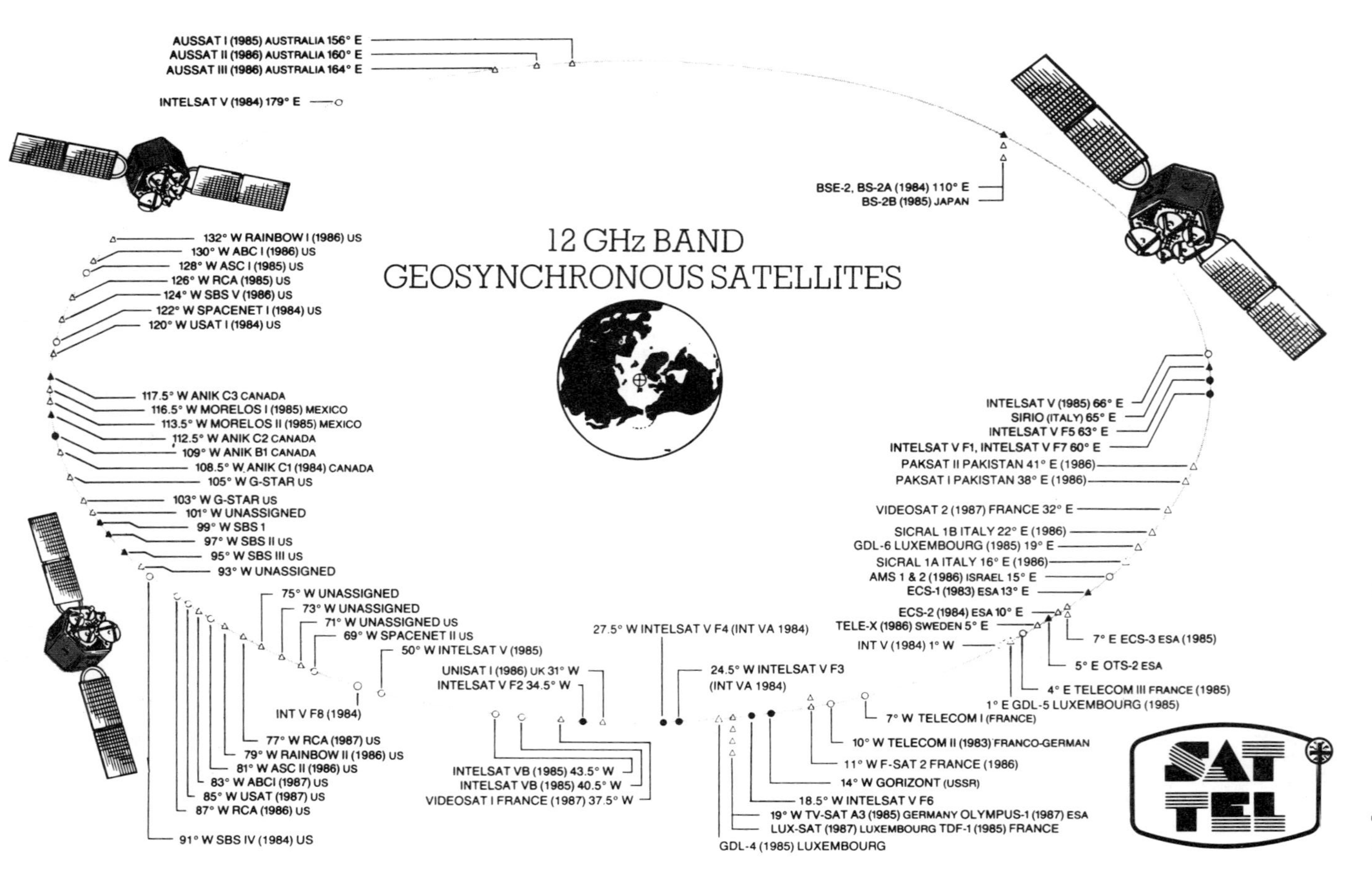
12 GHz BAND
GEOSYNCHRONOUS SATELLITES
AUSSAT I (1985) AUSTRALIA 156° E
AUSSAT II (1986) AUSTRALIA 160° E
AUSSAT III (1986) AUSTRALIA 164° E
INTELSAT V (1984) 179° E
BSE-2, BS-2A (1984) 110° E
BS-2B (1985) JAPAN
132° W RAINBOW I (1986) US
130° W ABC I (1986) US
128° W ASC I (1985) US
126° W RCA (1985) US
124° W SBS V (1986) US
122° W SPACENET I (1984) US
120° W USAT I (1984) US
117.5° W ANIK C3 CANADA
116.5° W MORELOS I (1985) MEXICO
113.5° W MORELOS II (1985) MEXICO
112.5° W ANIK C2 CANADA
109° W ANIK B1 CANADA
108.5° W ANIK C1 (1984) CANADA
105° W G-STAR US
103° W G-STAR US
101° W UNASSIGNED
99° W SBS 1
97° W SBS II US
95° W SBS III US
93° W UNASSIGNED
75° W UNASSIGNED
73° W UNASSIGNED
71° W UNASSIGNED US
69° W SPACENET II US
50° W INTELSAT V (1985)
INT V F8 (1984)
77° W RCA (1987) US
79° W RAINBOW II (1986) US
81° W ASC II (1986) US
83° W ABCI (1987) US
85° W USAT (1987) US
87° W RCA (1986) US
91° W SBS IV (1984) US
UNISAT I (1986) UK 31° W
INTELSAT V F2 34.5° W
27.5° W INTELSAT V F4 (INT VA 1984)
24.5° W INTELSAT V F3
(INT VA 1984)
INTELSAT VB (1985) 43.5° W
INTELSAT VB (1985) 40.5° W
VIDEOSAT I FRANCE (1987) 37.5° W
INTELSAT V (1985) 66° E
SIRIO (ITALY) 65° E
INTELSAT V F5 63° E
INTELSAT V F1, INTELSAT V F7 60° E
PAKSAT II PAKISTAN 41° E (1986)
PAKSAT I PAKISTAN 38° E (1986)
VIDEOSAT 2 (1987) FRANCE 32° E
SICRAL 1B ITALY 22° E (1986)
GDL-6 LUXEMBOURG (1985) 19° E
SICRAL 1A ITALY 16° E (1986)
AMS 1 & 2 (1986) ISRAEL 15° E
ECS-1 (1983) ESA 13° E
ECS-2 (1984) ESA 10° E
TELE-X (1986) SWEDEN 5° E
INT V (1984) 1° W
7° E ECS-3 ESA (1985)
5° E OTS-2 ESA
4° E TELECOM III FRANCE (1985)
1° E GDL-5 LUXEMBOURG (1985)
7° W TELECOM I (FRANCE)
10° W TELECOM II (1983) FRANCO-GERMAN
11° W F-SAT 2 FRANCE (1986)
14° W GORIZONT (USSR)
18.5° W INTELSAT V F6
19° W TV-SAT A3 (1985) GERMANY OLYMPUS-1 (1987) ESA
LUX-SAT (1987) LUXEMBOURG TDF-1 (1985) FRANCE
GDL-4 (1985) LUXEMBOURG
SAT TEL

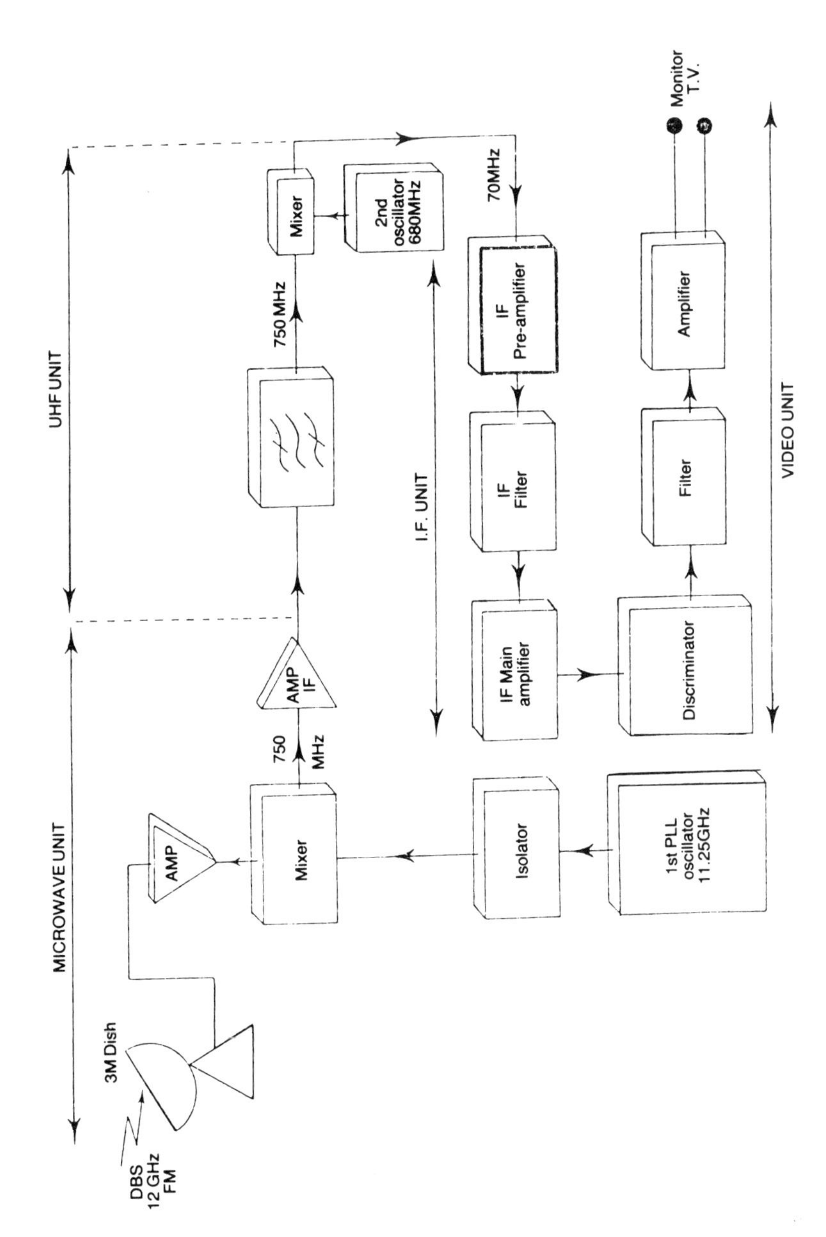

Fig. 98 – Circuit Configuration for professional Direct Broadcasting Satellite Ground Receiver.

IF main amplifier is to boost the IF in order that the discriminator functions properly.

The discriminator performs two functions: it suppresses amplitude variations and it detects the video in the FM carrier. Finally, the video is passed through a filter and amplifier to the TV monitor. The purpose of the low-pass filter is to separate the video information from any signals outside its passband, which is typically 5.5 MHz.

7.3 FIXED FREQUENCY DBS RECEIVER

The specification for a readily available receiving system made by Toshiba is shown in Fig. 99. An offset-feed antenna is used with a helical feed which the manufacturer claims will provide a wider than normal bandwidth and fewer production difficulties than a waveguide-type feed. It is interesting to see the claimed performance figures for the aerial assembly and the electronic units, the outdoor unit containing a GaAsFET amplifier and the indoor unit providing good sound quality from the pulse-code-modulated digital signals that the satellite radiates.

Specification for BS-2	
Receiving frequency:	11.7–12.0 GHz
Polarisation:	Right-hand circular polarisation
Antenna structure:	
Reflector:	Offset parabolic aperture diameter 1.0 m
Primary feed:	(a) Helical antenna, or
	(b) Horn antenna
Interface:	Coaxial or waveguide

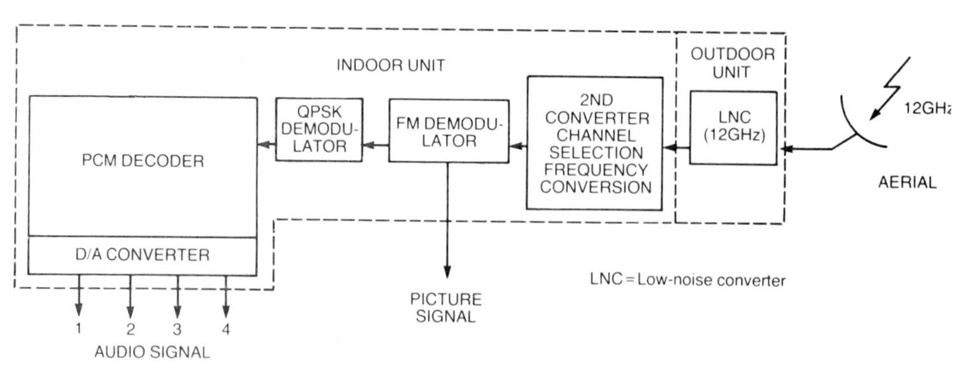

Fig. 99 – Toshiba DBS receiver configuration.

(In order to produce a low-cost antenna, it is more effective to use the coaxial interface).

Gain (CP):	More than 39.5 dBi
Side-lobe level:	WARC-BS Curve (Region 3)
Cross polarisation:	WARC-BS Curve (Region 3)
VSWR:	less than 1.3
Selectable channels:	8
IF bandwidth:	27 MHz
1st LO frequency:	10.678 GHz
1st IF:	1035–1322 MHz
2nd IF:	130 MHz
Picture S/N:	37 dB (unweighted at C/N = 14 dB)
Sound transmission method:	Sound PCM subcarrier (see Table 7.3)

Table 7.3 Sound transmission parameters

Item	Transmission mode A mode	B mode
Characteristic	Multi-channel mode	High quality sound mode
Audio signal bandwidth (kHz)	15	20
Sampling frequency (kHz)	32	48
Quantization bit	14–10 bit	16 bit
Companding	Near instantaneous companding	N/A
Audio channel	4	2
Independent bit	480 × bit	224 × bit
Bit rate (M bit/s)	2.048	
Error correction code	BCH	
Subcarrier frequency (MHz)	5.727272	
Subcarrier modulation method	QPSK	

Toshiba claim that their low-cost GaAsFET has a device noise figure of 1.6 dB, and they claim that an overall noise figure of 2.5 to 3 dB is feasible for mass-produced front ends. It is instructive to see Toshiba's graphs (Figs 100 and 101) showing the improvement in performance of both GaAsFETs and DBS receiver noise-figure/bandwidth characteristics over the last few years.

The need for careful design of FET amplifiers is stressed elsewhere in this book, together with some information about the calculation of the design parameters. A revolutionary development in receiver technology is about to take place that should considerable simplify the construction of low-cost front-ends. Some firms have developed a monolithic GaAs microwave integrated circuit which has a low-noise FET and all the ancilliary circuitry fabricated on just one chip. This design eliminates the need for careful

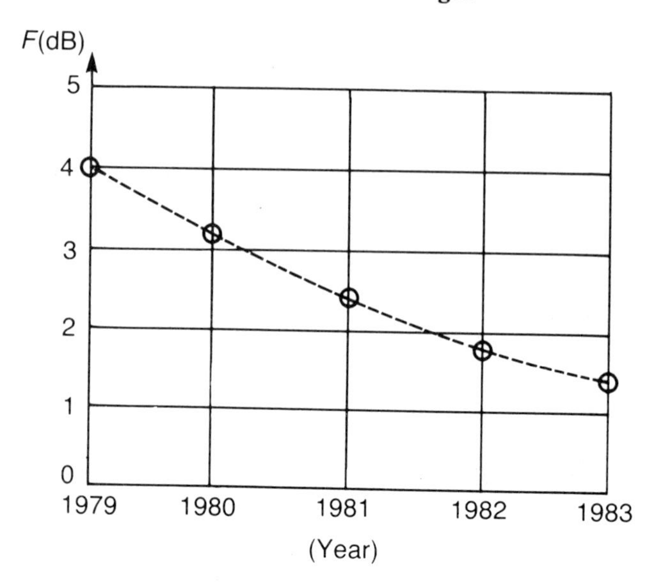

Fig. 100 – Noise performance of GaAsFET.

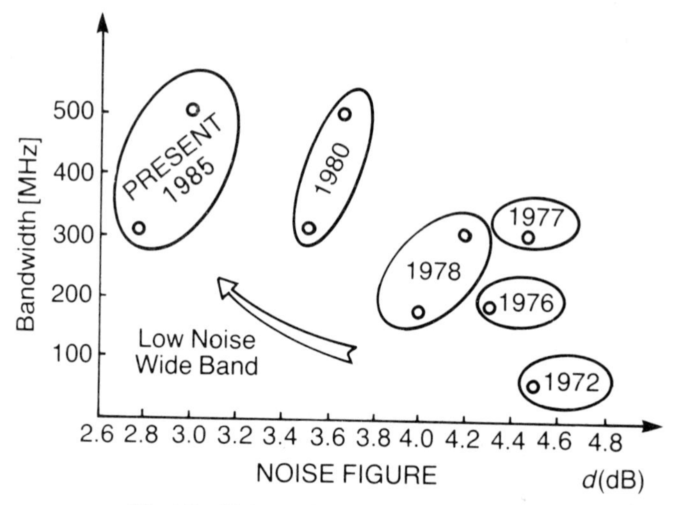

Fig. 101 – Noise performance of DBS receivers.

assembly by the user, and is obviously well suited for mass production. Provided that this device can be made in large quantities, it should be possible to make high-performance front ends for domestic receivers at low cost, which will help to ensure that satellite broadcasting is acceptable to a great many consumers.

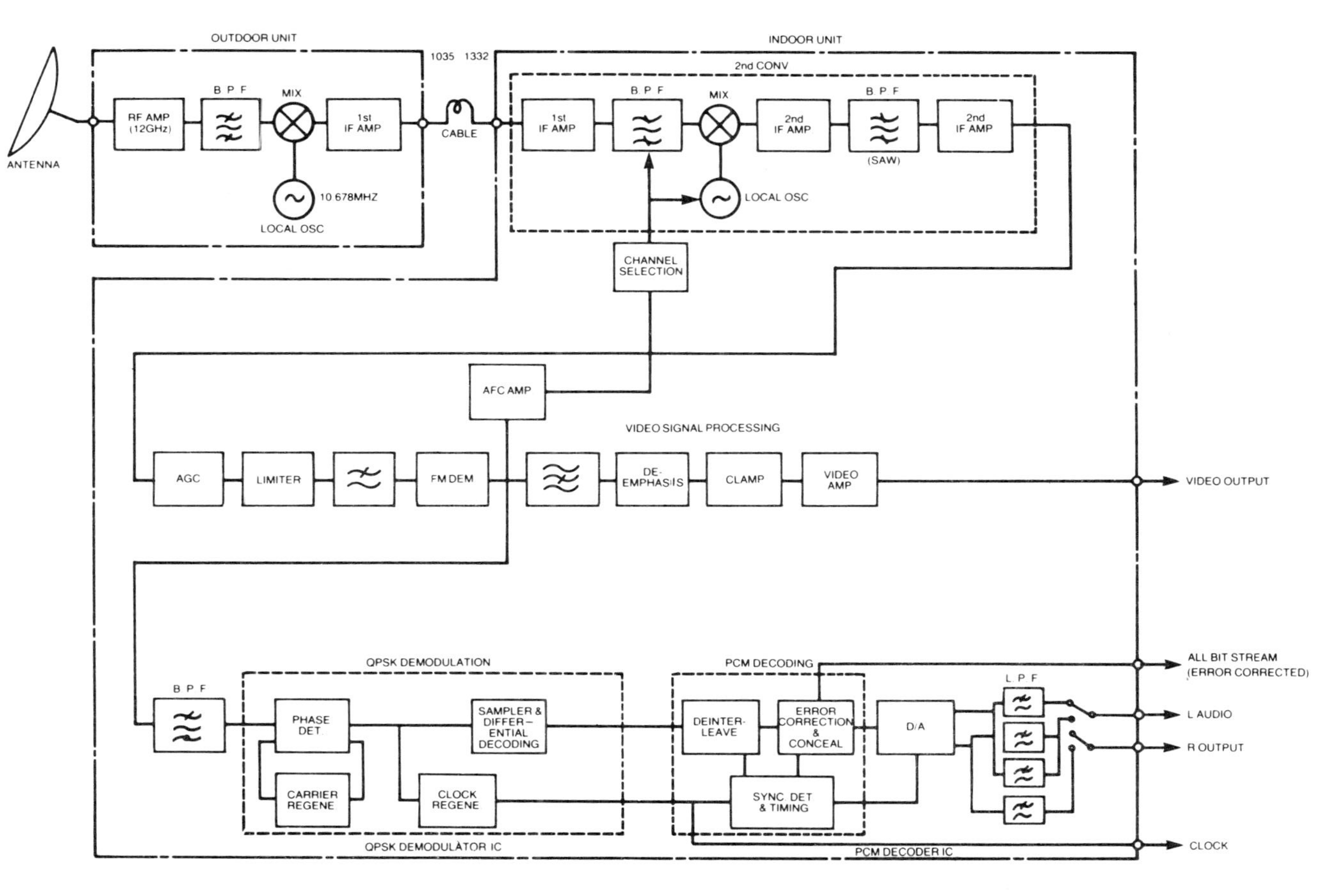

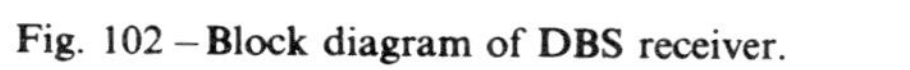
Fig. 102 – Block diagram of DBS receiver.

A more detailed block diagram of the Toshiba receiver design is shown in Fig. 102.

7.4 FLEXIBLE DBS RECEIVER

The use in Europe of ECS and INTELSAT V for satellite TV requires flexible DBS receiver design. As can be seen from Table 7.4, none of the channels is directly compatible. One way to provide flexibility is to use microprocessor controlled switching from one system to another as employed by SAT-TEL Space Communications of Northampton. Their low-noise converter is designed for an uncooled noise figure of 2.5 dB. The compact unit (Fig. 103) has sufficient gain to drive 50 m of coaxial cable. The unit employs packaged GaAsFETs which provide 27 dB gain at a noise figure of about 2 dB. The balanced mixer is driven by a dielectrically stabilised FET oscillator. The multistage IF transistor amplifier provides 30 dB gain over the frequency range 950–1750 MHz. A typical specification for the converter is given in Table 7.5. The low-noise converter (LNC) drives a video receiver whose block diagram is shown in Fig. 104. It covers the IF range of 950–1750 MHz. If multi-channel reception from one LNC is required, then a signal splitter must be employed. The video IF is down-converted to 137 MHz by means of harmonic conversion. The local oscillator is synthesised, and channel selection is performed with a microprocessor based system. The latter also controls the IF bandwidth, sound sub-carrier, and de-emphasis network. The coverage zones for ECS1 and INTELSAT V are shown in Fig. 105. The contours are marked in effective incident radiated power EIRP.

The following are examples of link budgets for a cable DBS receiver for ECS1 and INTELSAT V. The receiver, also known as television receive only (TVRO), is assumed to be positioned in Leeds.

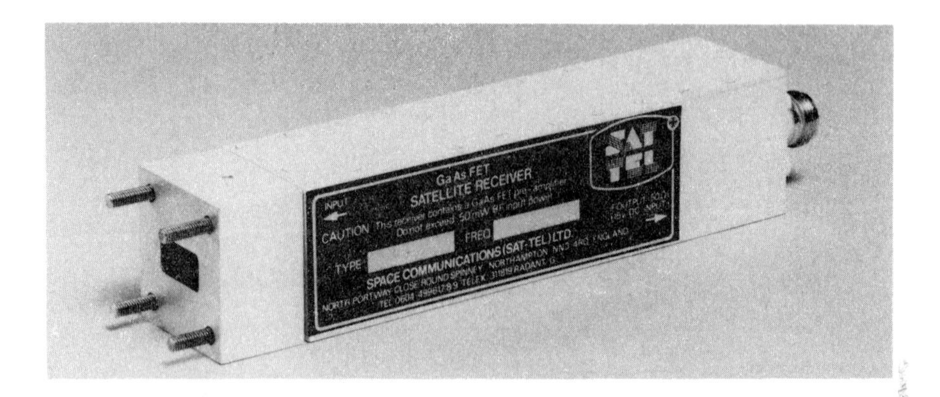

Fig. 103 – SAT-TEL space communications receiver.

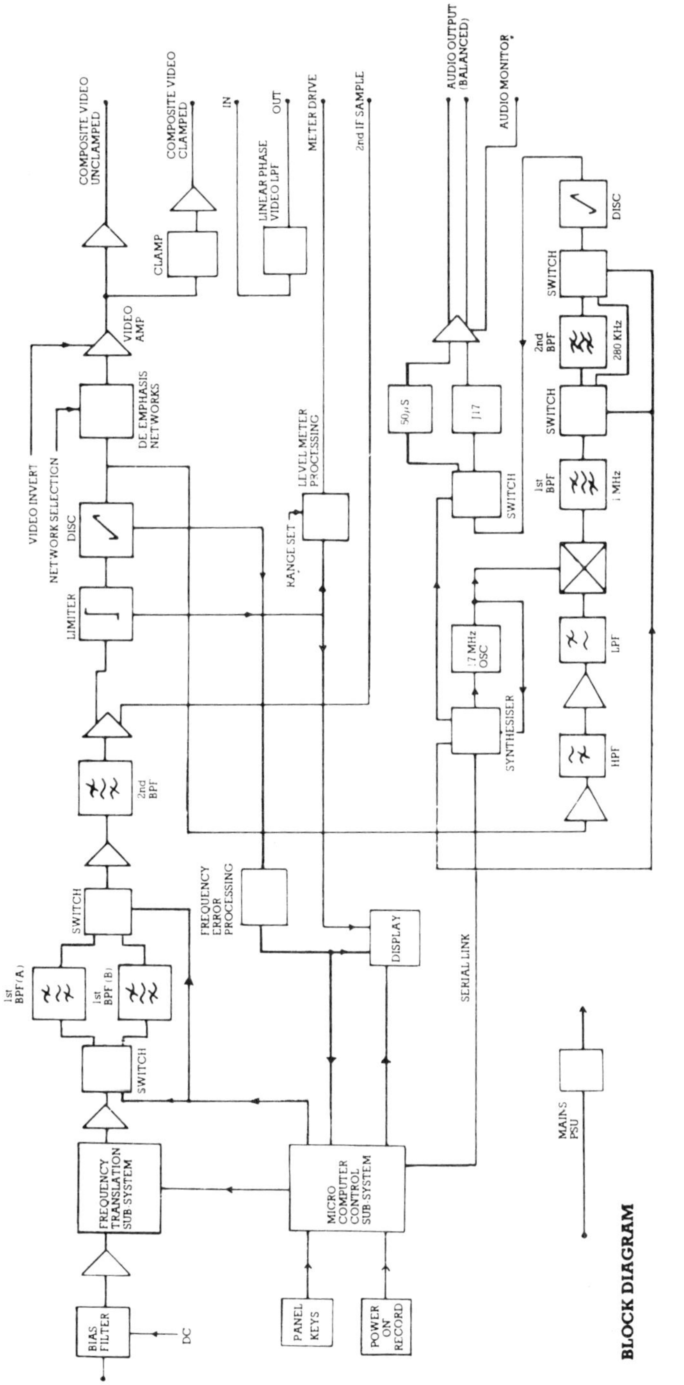

Fig. 104 – Video receiver block diagram.

Table 7.4 European satellite systems

Channel	deviation (MHz)	sub-carrier (MHz)	pre-emphasis (μs)
Music box	25	6.65	50
Rai	25	6.50	50
Sky	16	digital	—
Teleclub	25	6.50	75
Ten	20	6.65	50
TV5	25	6.65	J17

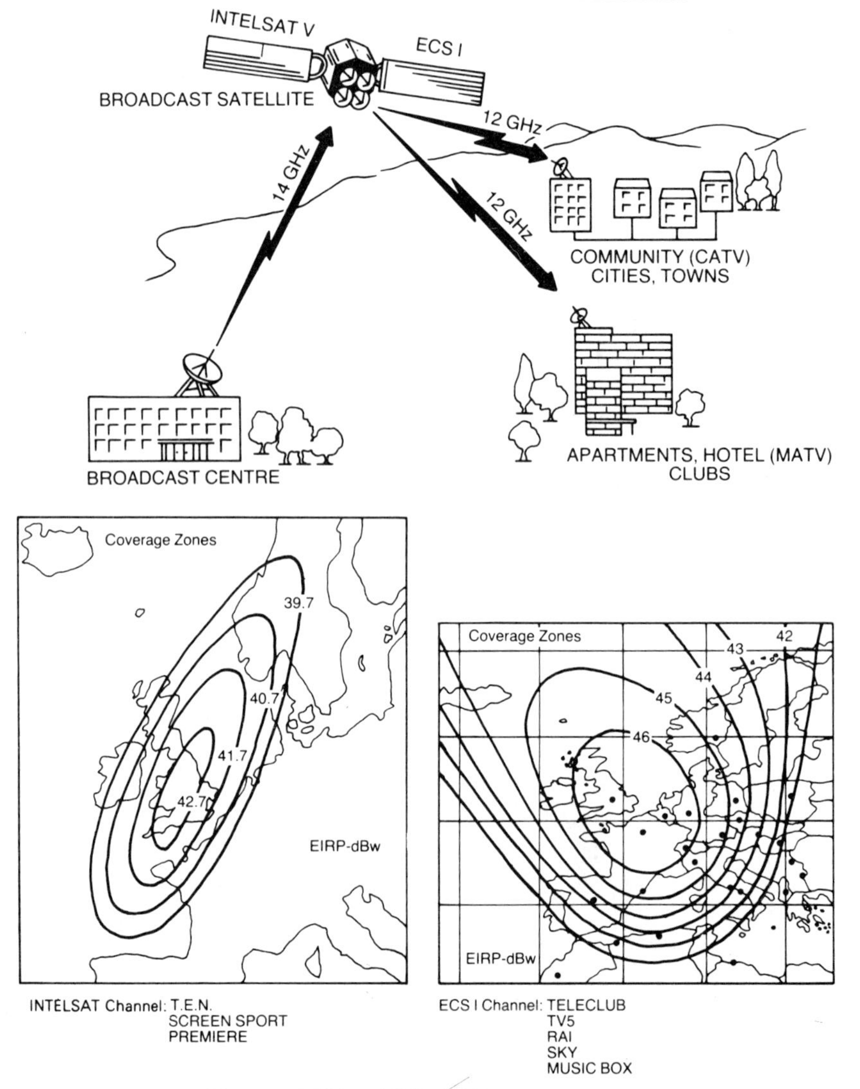

Fig. 105 – ECS1 and INTELSAT coverage zones.

Table 7.5 Specification of some SAT-TEL low-noise converters

Model	LNC HB3	LNC FB2
RF input Frequency range	11.45–11.8 GHz	10.95–11.7 GHz
IF output Frequency	0.95–1.3 GHz	0.95–1.75 GHz
Noise figure	3.0 dB typ (25°C)	25 dB typ (25°C) 2.2 dB typ (0°C)
LO frequency	10.5 ± 0.002 GHz	10.0 ± 0.002 GHz
RF to IF Conversion gain	35 dB typ	50 dB min
Gain flatness	± 2dB	± 2dB
Intercept point for 3rd order intermodulation	+ 10 dBm (min)	+ 15 dBm (min)
Input VSWR	2.1:1 max	1.25:1 max
Output VSWR	1.5:1	2:1 max
LO re-radiation	– 40 dBm	– 45 dBm
Image rejection	30 dB min	35 dB min
Operating temp range	– 30 to + 55°C	– 30 to + 55°C
Power supply requirements	12–25 V 200 mA	12–25 V 200 mA
Case size	140 × 55 × 50	175 × 40 × 40 mm
Weight	500 g approx	500 g approx
Connectors		
Input	WG 16 (17 opt)	WG 17
Output	N type 50 Ω	N type 50 Ω

REFERENCES

[1] T. C. Edwards, *Foundations for microstrip circuit design*. John Wiley and Sons, New York, 1981.

[2] W. Schilling, The real world of micromin substrates. Part 1 *Microwaves*, December 1968, pp. 52.

[3] Satellites for broadcasting. *IBA Technical Review*, No. 11, July 1978, pp. 31.

[4] S. Sando, A two-stage 3.7–4.2 GHz GaAsFET LNA. *Microwave Journal*, **24,** No. 10, Oct. 1981, pp. 107.

[5] S. J. Greenhalgh, Receivers evolving for TV by satellite. *Microwave Systems News*, **10,** No. 9, Sept. 1980, pp. 82.

[6] G. Galli *et al.*, Integration shrinks microwave front ends. *Microwave Systems News*, **10,** No. 9, Sept. 1980, pp. 119.

[7] M. G. Fornaciari, Integration simplifies TVRO design. *Microwave Systems News*, **11,** No. 3, March 1981, pp. 74.

[8] A. F. Podell, Monolithic microwave front ends find application niche. *Microwave Systems News*, **14,** No. 4, April 1984, pp. 48.

EXAMPLE OF LINK BUDGET FOR CABLE TV HEADEND RECEPTION FROM INTELSAT V

```
CHANNEL MUSIC BOX(UK)
SATELLITE:   ECS-F1 13 DEG E
CHANNEL FREQ(GHz):     11.67400
LINEAR POLARISATION-VERTICAL^
STEREO SOUND
MONO SOUND S/C(MHz):     6.65
R CH SOUND S/C(MHz):     7.02
L CH SOUND S/C(MHz):     7.20
PRE-EMPHASIS(uSec):    50
FM DEVIATION(MHz/V):    25.0
VIDEO BANDWIDTH(MHz)     5.5
CARSON BANDWIDTH(MHz): 36.0
TVRO LOCATION: BRUSSELS

TVRO LAT:                51.00 N
TVRO LONG:                4.50 E
TVRO ELEV:               31.05
TVRO AZIM:               10.89 E o S
DIST TO SAT(km):   38519.19

TVRO ANTENNA SIZE(m):        3.0
LNC NOISE FIG(dB):           2.4

LINK BUDGET
SATELLITE EIRP(dBW):         46.0
FREE SPACE LOSS(dB):        205.5
ATTENUATION DUE TO OXYGEN,WATER
VAPOUR&CLOUD(dB)              0.5
AERIAL GAIN(dB):             49.42
AERIAL EFF.(%) :             65.0
ANTENNA 3dB BEAMWIDTH(DEG)     .55
EFF.LOSS IN ANT.GAIN(dB):      .78
SIGNAL POWER(dBW):         -110.6
LNC NOISE TEMP(DEG. K):     211.47
ANTENNA NOISE TEMP.(DEG K):65.00
TOTAL SYSTEM NOISE TEMP:   276.47
SYSTEM(G/T) (dB):            25.00
SYSTEM NOISE (dBW):        -128.62
R/X NOISE BANDWIDTH(MHz):   36
CARRIER/NOISE(dB):           18.04
FM IMP FAC.(dB)              36.77
S/N(UNWEIGHTED)(dB):         43.16
S/N(WEIGHTED) (dB):          54.81

C/N(dB) 99.9% TIME:          15.54
S/N(W'TED)(dB) 99,9% TIME:  52.31
ASSUMES ESA RAIN MODEL
```

```
CHANNEL:RAI(ITALY)
SATELLITE:    ECS-F1 13 DEG E
CHANNEL FREQ(GHz):     11.00500
LINEAR POLARISATION-HORIZONTAL>
SOUND S/C(MHz):           6.70
PEAK DEV(kHz):           50
PRE-EMPHASIS(uSec):      50
FM DEVIATION(MHz/V):     25.0
VIDEO BANDWIDTH(MHz)      5.5
CARSON BANDWIDTH(MHz):  36.0
PRK 80 CHANNEL No:        21

TVRO LOCATION: NORTHAMPTON

TVRO LAT:                 52.00 N
TVRO LONG:                  .80 W
TVRO ELEV:                29.13
TVRO AZIM:                17.31 E o S
DIST TO SAT(km):       38689.98

TVRO ANTENNA SIZE(m):           2.0
LNC NOISE FIG(dB):              2.5

LINK BUDGET
SATELLITE EIRP(dBW):           45.0
FREE SPACE LOSS(dB):          205.0
ATTENUATION DUE TO OXYGEN,WATER
VAPOUR&CLOUD(dB)                 0.5
AERIAL GAIN(dB):                44.92
AERIAL EFF.(%)  :               56.5
ANTENNA 3dB BEAMWIDTH(DEG)        .92
EFF.LOSS IN ANT.GAIN(dB):         .28
SIGNAL POWER(dBW):           -115.6
LNC NOISE TEMP(DEG  K):       225.70
ANTENNA NOISE TEMP.(DEG K):65.00
TOTAL SYSTEM NOISE TEMP:   290.70
SYSTEM(G/T) (dB):               20.29
SYSTEM NOISE (dBW):          -128.40
R/X NOISE BANDWIDTH(MHz):   36
CARRIER/NOISE(dB):              12.80
FM IMP FAC.(dB)                 36.77
S/N(UNWEIGHTED)(dB):            37.92
S/N(WEIGHTED) (dB):             49.57


C/N(dB) 99.9% TIME:             10.30
S/N(W'TED)(dB) 99,9% TIME:  47.07
ASSUMES ESA RAIN MODEL
```

```
CAUTION!-MARGINALLY ENGINEERED
FOR URBAN CABLE SYSTEMS
SUITABLE FOR MATV SYSTEMS ONLY!
```

EXAMPLE OF LINK BUDGET FOR CABLE TV HEADEND RECEPTION FROM ECS 1

```
CHANNEL MUSIC BOX(UK)
SATELLITE:   ECS-F1 13 DEG E
CHANNEL FREQ (GHz):     11.67400
LINEAR POLARISATION-VERTICAL
STEREO SOUND
MONO SOUND S/C(MHz):       6.65
R CH SOUND S/C(MHz):       7.02
L CH SOUND S/C(MHz):       7.20
PRE-EMPHASIS(µSec):      50
FM DEVIATION(MHz/V):     25.0
VIDEO BANDWIDTH(MHz)      5.5
CARSON BANDWIDTH(MHz):   36.0
TVRO LOCATION: BRUSSELS

TVRO LAT                51.00 N
TVRO LONG:               4.50 E
TVRO ELEV:              31.05
TVRO AZIM:              10.39 E o S
DIST TO SAT(km):    38519.19

TVRO ANTENNA SIZE(m):          3.0
LNC NOISE FIG(dB):             2.4

LINK BUDGET
SATELLITE EIRP(dBW):           46.0
FREE SPACE LOSS(dB):          205.5
ATTENUATION DUE TO OXYGEN.WATER
VAPOUR&CLOUD(dB)                  0.5
AERIAL GAIN(dB):               49.42
AERIAL EFF.(%):                65.0
ANTENNA 3dB BEAMWIDTH(DEG)       .55
EFF.LOSS IN ANT.GAIN(dB):        .78
SIGNAL POWER(dBW):          -110.6
LNC NOISE TEMP(DEG. K):      211.47
ANTENNA NOISE TEMP (DEG K):  65.00
TOTAL SYSTEM NOISE TEMP:     276.47
SYSTEM(G/T) (dB):              25.00
SYSTEM NOISE (dBW):         -128.62
R/X NOISE BANDWIDTH(MHz):    36
CARRIER/NOISE(dB):             18.04
FM IMP FAC.(dB)                36.77
S/N(UNWEIGHTED)(dB):           43.16
S/N(WEIGHTED) (dB):            54.81
```

```
C/N(dB) 99.9% TIME:           15.54
S/N(W'TED)(dB) 99,9% TIME: 52.31
ASSUMES ESA RAIN MODEL

CHANNEL:SKY(UK)
SATELLITE:    ECS-F1 13 DEG E
CHANNEL FREQ(GHz):     11.65000
LINEAR POLARISATION-HORIZONTAL)
ENCRYPTED SOUND
FM DEVIATION(MHz/V):    18.0
VIDEO BANDWIDTH(MHz)     5.5
CARSON BANDWIDTH(MHz): 29.0
TVRO LOCATION: BRUSSELS

TVRO LAT                 51.00 N
TVRO LONG                 4.50 E
TVRO ELEV                31.05
TVRO AZIM                10.29 E o S
DIST TO SAT(km):  38519.13

TVRO ANTENNA SIZE(m):         3.7
LNC NOISE FIG(dB):            2.9

LINK BUDGET
SATELLITE EIRP(dBW):          46.0
FREE SPACE LOSS(dB):         205.5
ATTENUATION DUE TO OXYGEN,WATER
VAPOUR&CLOUD(dB)                0.5
AERIAL GAIN(dB):              50.82
AERIAL EFF.(%) :              59.2
ANTENNA 3dB BEAMWIDTH(DEG)     .47
EFF.LOSS IN ANT.GAIN(dB):    1.08
SIGNAL POWER(dBW):         -109.2
LNC NOISE TEMP(DEG  K):    272.76
ANTENNA NOISE TEMP.(DEG K):65.00
TOTAL SYSTEM NOISE TEMP:  337.76
SYSTEM(G/T) (dB):             25.53
SYSTEM NOISE (dBW):        -128.54
R/X NOISE BANDWIDTH(MHz):  30
CARRIER/NOISE(dB):            19.38
FM IMP FAC.(dB)               33.12
S/N(UNWEIGHTED)(dB):          40.86
S/N(WEIGHTED) (dB):           52.50


C/N(dB) 99.9% TIME:           16.88
S/N(W'TED)(dB) 99,9% TIME: 50.00
ASSUMES ESA RAIN MODEL
```

```
  10 REM BORESIGHT
  20 INPUT "longitude of RX ?",l
r
  30 INPUT "latitude of RX ?",ph
ir
  40 INPUT "longitude of geostat
ionary satellite ?",ls
  45 INPUT "semi-beamw. ?",delta
  47 INPUT "frequency (GHz)",f
  50 LET h=35779
  60 LET l=ls-lr
  70 LET a=COS (phir*PI/180)*COS
 (l*PI/180)
  80 LET beta=(ACS a)*180/PI
  90 LET sigma=6386/(6385+35780)
 100 LET b=TAN (l*PI/180)/SIN (p
hir*PI/180)
 110 LET az=((ATN b)*180/PI)+180
 120 LET c=(COS (beta*PI/180)-si
gma)/SIN (beta*PI/180)
 130 LET el=(ATN c)*180/PI
 140 LET d=35779*SQR (1+.41999*(
1-COS (beta*PI/180)))
 141 REM calculation of free spa
ce pathloss "a0"
 143 LET x=(3*(10↑10)/(f*(10↑9)*
4*PI*d*10↑5))
 144 LET y=x↑2
 145 LET a0=10*(LN y/LN 10)
 168 REM satellite coverage
 170 LET a1=d*TAN ((delta*PI/180
))/SIN ((el*PI/180)-(delta*PI/18
0))
 175 LET a2=d*TAN ((delta*PI/180
))/SIN ((el*PI/180)+(delta*PI/18
0))
 180 LET E1=a1+(((a1*a1)/2)/(638
6*TAN ((el*PI/180)-(delta*PI/180
))-a1))
 185 LET E2=a2-(((a2*a2)/2)/(638
6*TAN ((el*PI/180)+(delta*PI/180
))+a2))
 190 LET E3=d*TAN (delta*PI/180)
 195 PRINT "BORESIGHT calculatio
ns"
 196 PRINT
 200 PRINT "l=ls-lr  (°)",l
 220 PRINT "azimuth (°)",az
 230 PRINT "elevation (°)",el
 240 PRINT "pathl'gth d(km)",d
 250 PRINT "frequency(GHz)",f
 251 PRINT
 252 PRINT "free space loss",a0;
" dB"
 260 PRINT
 270 PRINT "E1",E1
 280 PRINT
 290 PRINT "E2",E2
 300 PRINT
 310 PRINT "E3",E3
 315 PRINT
 320 PRINT "delta",delta
 321 PRINT
 322 PRINT
```

```
 323 PRINT "If you want a plot o
f the coverage area then RUN 333
"
 330 STOP
 333 REM coverage area
 335 PLOT 152,25: DRAW -65,65,-P
I
 336 PLOT 150,20: DRAW 40,80,PI/
10
 340 PLOT 80,80: DRAW 60,55,-PI/
5
 350 PLOT 190,100: DRAW -50,30,P
I
 360 PLOT 90,20: DRAW 90,120
 370 PLOT 155,30: DRAW -65,65
 380 PRINT AT 7,19;"E1"
 390 PRINT AT 11,12;"E3"
 400 PRINT AT 16,17;"E3"
 410 PRINT AT 16,12;"E2"
 411 PLOT 122,60: DRAW 0,90
 412 PRINT AT 2,15;"N"
 420 PRINT AT 1,1;"DBS coverage
"
 421 PRINT AT 2,1;"area"
 426 PRINT AT 20,0;"Want the lon
gitude and latitude of some citi
es?Then RUN 444"
 427 STOP
 445 PRINT "Longitudes (LON) and
 latitudes  (LAT) of some cities
"
 450 PRINT
 455 PRINT "                   LON
    LAT"
 456 PRINT
 457 PRINT
 460 PRINT "Belfast          5.55W
    54.35N"
 461 PRINT
 465 PRINT "London           0.10W
    51.30N"
 466 PRINT
 470 PRINT "Koln             6.57E
    50.56N"
 471 PRINT
 475 PRINT "Newcastle        1.35W
    54.59N"
 476 PRINT
 478 PRINT "Leverkusen       6.59E
    51.02N"
 479 PRINT
 480 PRINT "Capetown        18.28E
    33.56S"
 481 PRINT
 482 PRINT "Roorkee         77.54E
    29.51N"
 483 PRINT
 484 PRINT "Leningrad       30.25E
    59.55N"
 485 PRINT
 486 PRINT "Leeds            1.35W
    53.50N"
 500 STOP
```

(LAT) of some cities

	LON	LAT
Belfast	5.55W	54.35N
London	0.10W	51.30N
Koln	6.57E	50.56N
Newcastle	1.35W	54.59N
Leverkusen	6.59E	51.02N
Capetown	18.23E	33.56S
Roorkee	77.54E	29.51N
Leningrad	30.25E	59.55N
Leeds	1.35W	53.50N

```
   1 REM STRIP 1
   5 REM This programme calculst
es microstrip width and length u
sing a programme in BASIC.

  10 LET PII=3.1415
  20 INPUT "Z0 = ?",Z0
  30 INPUT "ER = ?",ER
  39 REM HI=thickness substrate
  40 INPUT "HEIGHT = ?",HI
  50 INPUT "FREQ(GHz) = ?",F
  60 LET H=Z0*SQR (2*(ER+1))/119
.9+(ER-1)/(ER+1)/2*(LN (PII/2)+L
N (4/PII)/ER)
  80 LET RATIO=1/(EXP (H)/8-1/(4
*EXP (H)))
  90 GO TO 120
 100 LET D=59.95*PII*PII/Z0/SQR
(ER)
 110 LET RATIO=2/PII*(D-1-LN (2*
D-1))+(ER-1)/PII/ER*(LN (D-1)+0.
293-0.517/ER )
 130  LET RADD=1/(1-(ER-1)/(ER+1
)/2/H*(LN (PII/2)+LN (4/PII)/ER)
*(LN (PII/2)+LN (4/PII)/ER))
 150 PRINT
 160 LET W=RATIO*HI
 165 LET EF=((ER+1)/2)+((ER-1)/2
)/SQR (1+(12*HI/W))
 169 REM LO=freespace wavelength
 170 LET LO=300/F
 179 REM LEFF=effective waveleng
th
 180 LET LEFF=LO/SQR (EF)
 190 LET LQ=LEFF/4
 200 LET LH=LEFF/2
 210 PRINT
 220 PRINT "W/H=";RATIO
 221 PRINT
```

```
230 PRINT "given H=";HI;"mm thus
W=";W;"mm"
240 PRINT
250 PRINT "Freespace wavelength
=";LO;"mm"
260 PRINT
270 PRINT "EEFF = ";EF
280 PRINT
290 PRINT "eff wavelength =";LE
F;"mm"
291 PRINT
300 PRINT "eff halfwave=";LH;"m
m"
301 PRINT
310 PRINT "eff quarterwave=";LQ
;"mm"
311 PRINT
312 PRINT
313 INPUT "would you like anoth
er frequency? Then enter Y",D$
314 IF D$="Y" THEN GO TO 50
315 PRINT
320 PRINT "to obtain amplifier
layout enter RUN 330"
330 STOP
340 INPUT "l1",l1
341 INPUT "l2",l2
342 INPUT "l3",l3
343 INPUT "l4",l4
345 PRINT "AMPLIFIER LAYOUT"
347 CIRCLE 125,120,13
348 PLOT 122,113: DRAW 0,15
349 PLOT 110,115: DRAW 0,10
350 PLOT 140,115: DRAW 0,10
351 PLOT 60,115: DRAW 50,0
352 PLOT 190,115: DRAW -50,0
353 PLOT 60,115: DRAW 0,10
354 PLOT 190,115: DRAW 0,10
355 PLOT 60,125: DRAW 50,0
356 PLOT 140,125: DRAW 50,0
360 REM length l2,l3 completed
361 PLOT 55,90: DRAW -10,0
362 PLOT 195,90: DRAW 10,0
363 PLOT 55,90: DRAW 0,-50
364 PLOT 45,90: DRAW 0,-50
365 PLOT 45,40: DRAW 10,0
366 PLOT 195,40: DRAW 10,0
367 PLOT 195,40: DRAW 0,50
368 PLOT 205,40: DRAW 0,50
400 REM length l1,l4 completed
401 PLOT 60,120: DRAW -20,0
402 PLOT 190,120: DRAW 20,0
403 PLOT 50,120: DRAW 0,-30
404 PLOT 200,120: DRAW 0,-30
405 PLOT 40,115: DRAW 0,10
406 PLOT 35,115: DRAW 0,10
407 PLOT 35,120: DRAW -20,0
408 PLOT 215,120: DRAW 20,0
409 PLOT 210,115: DRAW 0,10
410 PLOT 125,95: DRAW 10,0
411 PLOT 215,115: DRAW 0,10
412 REM capacitors finished.now
```

```
complete transistor
415 PLOT 122,120: DRAW -5,0
416 PLOT 122,120: DRAW 8,-8
417 PLOT 130,111: DRAW 0,-15
418 PLOT 125,96: DRAW 10,0
419 PLOT 125,94: DRAW 10,0
420 PLOT 125,120: DRAW 5,5
421 PLOT 131,126: DRAW 3,0
425 PRINT
426 PRINT
430 PRINT "                    TR"
432 PRINT "    C in    l2          l
3    C out"
434 PRINT
435 PRINT
436 PRINT
437 PRINT
438 PRINT
439 PRINT
440 PRINT
441 PRINT "    l1
 l4"
444 PRINT
445 PRINT
446 PRINT
447 PRINT
448 PRINT
450 PRINT "l1 =";l1;"mm"
452 PRINT "l2 =";l2;"mm"
454 PRINT "l3 =";l3;"mm"
455 PRINT "l4 =";l4;"mm"
500 STOP
```

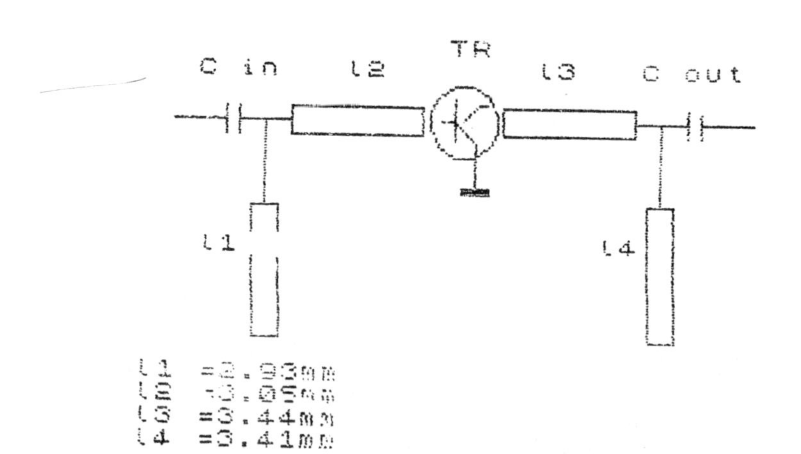

Index